NEVER SAY
IT'S JUST A DANDELION

NEVER SAY
IT'S JUST A DANDELION

125 Wonderful Common Plants for Walkers and Walk Leaders

by Hilary Hopkins

Jewelweed Books
Cambridge, Massachusetts

ISBN 0-9711048-0-8

Library of Congress Control Number 2001 135422

Printed in the United States of America
Manufactured by BookMasters, Inc.
2541 Ashland Road, P. O. Box 2139
Mansfield Ohio 44905

Jewelweed Books
30 Winslow Street
Cambridge, Massachusetts 02138
617-491-8369
email: JewelweedBooks@aol.com

Hopkins, Hilary 1938-
Never Say It's Just A Dandelion:125 Wonderful
Common Plants for Walkers and Walk Leaders/by
Hilary Hopkins
        Includes bibliography and index.
        LCCN: 2001 135422
        ISBN: 0-9711048-0-8

Thank you, precious Mother

Thank you, dearest John, my Elephant

INTRODUCTION

At the beginning of a walk, people often ask the leader what wildlife they might expect to see. By "wildlife" they almost always mean animals. But the plants are also wild life, as marvelous and awesome as the animals. There are wonderful plant delights to be found hidden in plain view on your local walk, in the Dandelion, the Goldenrod, in Dock and Oak.

The general range of this book is Eastern Massachusetts, but the plants included are so common (that is, abundant!) that the book can serve as a useful reference in other places, too.

There are seven entries for each plant..

About the names tells the meanings and sometimes the origins of both common and scientific names. There are some wonderful stories to be found here. Many plants were named by the Ancients—just imagine old Theophrastus (372-287 B. C.) standing by a pond over two thousand years ago and contemplating the same Duckweed found in your local pond!

Memory aid helps you remember the name of the plant, usually by linking its name with something about its appearance.

When in bloom helps you to find plants through the seasons.

Interesting growth habits invites you to notice some of the wondrous complexities and beauties of each plant.

Warnings and uses calls your attention first to anything harmful about each plant. I have then selected a few of the more striking uses to which the plant has been put by ingenious humans. **Please note that no claims whatsoever are made for the safety or efficacy of any of the food or medicinal uses described.**

"Stories" includes a miscellany of intriguing observations and lore.

Do this offers invitations to engage with the plant—to smell or touch it, look with a magnifier, count things, measure things...

Finally, opposite each plant entry is a whole page for you to do as you like with, to make notes or sketches or whatever.

Enjoy yourself! And never say "It's just a dandelion." For wonders are commonplace, and the common is wonderful.

Hilary Hopkins

125 Wonderful Common Plants

| | **Smooth Alder** |
	Alnus serrulata
About the names	• <u>Alder</u> from an Old English word for this tree; <u>smooth</u> describes the bark, especially as compared to that of the Speckled Alder, which this tree closely resembles. • <u>Alnus</u>: Latin word for the genus; <u>serrulata</u> means "edged with small teeth," referring to the leave s. • Also called Common Alder.
Memory aid	• Think: smooth cat[kin] paw -lder. • Note that Smooth and Speckled Alder are the only common shrubby plants with "cones" and catkins on them all winter; and the only common Alders to have reddish buds on short stalks.
When in bloom	• Large drooping male and smaller, <u>non-drooping</u> female catkins form in fall; flowers open in February-May and pollen is released; fruit ("cones") forms and seeds released in fall; catkins of both sexes, as well as female "cones," remain on tree through winter.
Interesting growth habits	• Bacteria living in root nodules can fix nitrogen from atmosphere, like legumes. • Leaves are extremely high in nitrogen, and therefore decay very rapidly. • Frequently hybridizes with Speckled Alder. • Male catkins expand to several times their original length when they flower. • Alders need oxygenated water, so their presence near a stream indicates good water quality.
Warnings, uses	• In a pinch, grind up the inner bark to make flour. • A field of Alders can add as much as 140 pounds of nitrogen per acre of ground.
"Stories"	• Alder is sometimes host to black alder jelly, a fungus which looks like black paint on a dead Alder branch, until it rains, when it expands and looks like lumpy black jelly.
Do this	• In spring, use a magnifier to observe the tiny flowers; look for tiny reddish hairs.

For your sketches, notes, observations

	Speckled Alder *Alnus rugosa*
About the ***names***	• <u>Alder</u> from an Old English word describing this tree; <u>speckled</u> describes the bark. • <u>Alnus</u>: Latin word for the genus; <u>rugosa</u> means "wrinkled," referring to appearance of leaves, with their network of sunken veins. • Also called Tag Alder, Black Alder.
Memory ***aid***	• Think: speckled cat[kin] paw-lder. • Note that Speckled and Smooth Alder are the only shrubby plants to have both "cones" and catkins on them all year; and the only Alders with reddish buds on short stalks.
When in ***bloom***	• Large drooping male and smaller <u>drooping</u> female catkins form in fall; flowers open in March-May and pollen is released; fruit ("cones") forms and seeds released in fall; catkins of both sexes, as well as female "cones," remain on tree through winter.
Interesting ***growth*** ***habits***	• Able to fix nitrogen, like a legume, via bacteria living in nodules on roots. • Leaves also very rich in nitrogen; therefore decays in leaf litter in about a year. • Frequently hybridizes with Smooth Alder. • Elongated speckles are lenticels, holes into the bark for gas exchange.
Warnings, ***uses***	• To replenish nitrogen-poor soil: a field of Alders can add as much as 140 pounds of nitrogen per acre. • Woodcocks favor Alder thickets for nesting (these birds eat worms; see below about worms and Alders). • Dense mats of roots help prevent erosion. • Chippewa Indians mixed scrapings from the roots with ground-up bumblebees, and administered two tablespoons to women in difficult childbirth.
"Stories"	• To rid your woodlot of spreading Alders, be sure to cut them in the black moon of August—that will ensure there will be no resprouting. • Earthworms, that useful animal, prefer the decaying leaves of Alders for food, because of their high nitrogen and sugar content.
Do this	• Pick a "cone" and shake it gently into your hand to see the tiny wind-carried seeds.

For your sketches, notes, observations

CAUTION	**Rue Anemone** *Anemonella thalictroides*
About the names	• <u>Rue</u> from an ancient Peloponnesian word for a plant with leaves shaped like some of those in this species; <u>Anemone</u> and <u>Anemonella</u> come from Greek meaning "the wind," applied to these flowers possibly because they are supposed to open only when stirred by spring winds, or because they move in the slightest breeze; there is also a story that the name comes from a name for Adonis, from whose blood some Anemones grew. In any case, the names are all very old. • <u>Thalictroides</u> comes from the name Thalia, the Greek Muse who presided over pastoral poetry; the suffix <u>–oides</u> means "resembling."
Memory aid	• Lots of 3's, like the 3 letters in r-u-e: 3-lobed leaves in 3 groups of 3. Flowers are delicate: visualize an <u>anemo</u>-meter, measuring the wind that gently waves them.
When in bloom	• March/April-May/June.
Interesting growth habits	• A similar flower, Wood Anemone, closes on cloudy days, but Rue Anemone stays open. • Stem and leaf stalks are like black wire. • The "flower" actually has no petals; what look like petals are actually sepals, a different plant part. • A spring ephemeral, blooming early, before trees leaf out, and disappearing without a trace until next spring.
Warnings and uses	• CAUTION Possibly toxic; as with any plant, do not experiment!
"Stories"	• A preparation from the roots used to be used to treat hemorrhoids. We repeat: do not experiment! • One source says this plant is "easily cultivated in wildflower gardens." Of course one would not dig it up from its woodland home!
Do this	• Try drawing or outlining the pleasing leaf shape; this will help in recalling "rue" leaves the next time.

For your sketches, notes, observations

	Wood Anemone
	Anemone quinquefolia
About the names	• Origin of <u>Anemone</u> not clear: possibly from Greek meaning "wind," since, according to one speculation, the flowers are as fleeting as the wind; <u>Wood</u> for its habitat. • <u>Quinque-folia</u> means "five-leaved," referring to leaves in five parts. • Also called Drops-of-Snow, Granny's-Nightcap, Little-Buffalo-Medicine, Snowboys, Windflower.
Memory aid	• A tough one. Like the sea-dwelling anemone animal, the wood-dwelling Anemone flower closes its "petals" when conditions aren't right.
When in bloom	• April-June.
Interesting growth habits	• May grow in small colonies. • "Petals" are really sepals, a different plant part; white on top, may be pinkish underneath. • Just one flower on each stem. • Flower closes on cloudy days. • Spring ephemeral, blooming early before tree leaves shade it out, and after two weeks, disappears without a trace until next season.
Warnings, uses	• "It is said to be extremely acrid—even small doses producing a great disturbance of the stomach; employed…in fevers, gout, and rheumatism, and …in removing corns…" (1865, Dr. Porcher, a surgeon in the Confederate Army)
"Stories"	• Used by the Chinese in funeral rites, symbolizing death. • Romans picked first one of the season for fever prevention; a certain cure was to pick one, say, "I gather this against all diseases," and tie it around the sick person's neck.
Do this	• Check under the "petals" to see if they are pink instead of white. • Is it a cloudy day? See if you can find some closed Wood Anemones.

For your sketches, notes, observations

POISONOUS	**Arrow-Arum**
	Peltandra virginica

About the names	• <u>Arrow-Arum</u>: a member of the Arum family (like Jack-in-the-Pulpit), with arrow-shaped leaves. • <u>Peltandra</u>: from two Greek words meaning "small shield" and "stamen," referring to the shapes of the flower parts; <u>virginica</u> means "from Virginia." • Also called Wild Calla or Tuckahoe.
Memory aid	• <u>Arrow</u>-shaped leaves with <u>a-rim</u> [arum] vein around their edges.
When in bloom	• May-July.
Interesting growth habits	• Has a spathe (the enclosing "leaf") and spadix (rod-like flower) arrangement like Jack-in-the-Pulpit; long pointed spathe almost completely encloses spadix (very weird-looking). • Green female flowers are at base of spadix, white male flowers above. • Likes shallow waterways, swamps, and bogs. • Leaves have three prominent veins plus a vein along the margin. • As the fruits mature, the spathe swells and bends down; sometimes its top drops off and seeds may be seen in a kind of jelly. If it doesn't drop off, the tip of the spathe may "drill" into mud and thus plant seeds.
Warnings, uses	• CAUTION Contains calcium oxalate crystals in all parts; if eaten will cause burns and irritation and actually cut tiny wounds. • Eaten as a cooked vegetable, or roots ground as flour, BUT ONLY after thorough drying!
"Stories"	• Descriptions of Native American methods of preparing this for food are quite heroic: wrap in oak leaves and ferns and cook in a pit for 24 hours, or boil steadily for eight or nine hours—the problem is that it still might "prickle and torment the throat extremely," or "make a man for the time frantick or extremely sick."
Do this	• See if you can find Arrowhead, Pickerelweed, and Arrow-Arum growing near each other; compare the three and learn to spot the differences in leaves and in flowers.

For your sketches, notes, observations

CAUTION	**Arrowhead**
	Sagittaria latifolia
About the names	• <u>Arrowhead</u>: exactly descriptive of the leaves. • <u>Sagittaria</u> also means "arrowhead"; <u>latifolia</u> means "broad-leaved." • Also called Duck-Potato.
Memory aid	• Flowers in three's, each with three petals, like the three syllables of its name, which perfectly describes the leaves.
When in bloom	• June-September.
Interesting growth habits	• May be pollinated by insects or even snails. • Lives in shallow water near the edge of the pond; an emergent, with roots in underwater soil and stems, leaves, and flowers well above water-line. • Note that leaves have parallel veins. • Flowers and leaves are on separate stems; female flowers low on stem, showy male flowers above.
Warnings, uses	• CAUTION May cause skin irritation. • Tubers, the size of golf balls, are edible when cooked like potatoes; loosen them with a rake and they will float to the surface. • Native Americans used a poultice of the leaves to stop mothers' milk production.
"Stories"	• So efficient at moving pond water through leaves and respiring it that they have been eliminated at some reservoirs because they can significantly lower water level. • Native Americans would wade through the water, feeling the "potatoes" with their toes in order to find and dig them out. • Sometimes muskrat lodges were raided to take the cache of "potatoes." • Lewis and Clark observed, in 1806, that the "potatoes" were a staple food and even a trading item on the Columbia River.
Do this	• Use binoculars to take a close look at the lovely flowers, which somehow seem orchid-like.

For your sketches, notes, observations

| | **Arrowwood** |
	Viburnum dentatum
About the names	• <u>Arrowwood</u>: long, straight, slender branches. • <u>Viburnum</u>: the Latin name for the Wayfaring-Man's-Tree, a member of this genus; <u>dentatum</u> means "with outward-pointing teeth," referring to the leaves.
Memory aid	• Very large, regular teeth threaten like a fortress of arrowheads around edge of leaf.
When in bloom	• May-July; fruits July-September.
Interesting growth habits	• Opposite-branching hairy ridged twigs. • The shrub has many long stems growing from the same place. • Cluster of small white flowers grows in an umbrella shape, followed by purplish-black or blue-gray fruits in the same form, thirty or more of them in a cluster. • Large flat seeds have a deep, narrow groove. • Likes to live on the edges of swamps. • Leaves turn bronze or red in fall.
Warnings, uses	• Chipmunks and ruffed grouse enjoy the fruits. • Native Americans are said to have used the very regular stems to make arrows.
"Stories"	• Viburnums are "the classical shrubs of the woods" (Donald Stokes).
Do this	• If it is in fruit, open one to check for the groove on the seed. • Use a magnifier to see the tiny hairs on the leaf veins.

For your sketches, notes, observations

	Quaking Aspen *Populus tremuloides*
About the names	• <u>Aspen</u> is possibly derived from an Old English word meaning "trembling"; "<u>quaking</u>" refers to a habit of the leaves. • <u>Populus</u>: ancient name for Poplar; <u>tremuloides</u> meaning "trembling." • Also called Trembling Aspen, Popple.
Memory aid	• Leaves flutter or quake in the slightest breeze.
When in bloom	• Early spring, before leaves appear; fruit (catkins) mature in late spring.
Interesting growth habits	• Leaf stalks are flat (like ribbons), which allows leaves to flutter easily. • Among first trees to appear in burned areas, since seeds can only germinate on bare soil. • Spread by an underground cloning root system; a stand may include over a hundred individuals, each genetically identical. • May be two feet tall at two years. • Cotton-like seeds, incredibly light (about 3 million per pound!), are wind-dispersed in early spring, thus taking advantage of still-bare ground and open canopy. • Young winter bark is pale, pale green.
Warnings and uses	• Beavers love the twigs and bark, both for building and eating; many other mammals and birds eat all parts of Aspen trees. • Bark contains salicin, related to active ingredient of aspirin; used by Native Americans to treat fever, pain, infection. • Buds were boiled with fat to make salve for sore nostrils. • Drink made from roots was used to prevent premature childbirth. • Wood used for paper pulp, matches, crates.
"Stories"	• A stand in Minnesota has 50,000 trees and covers 200 acres; aged at 8,000 years, making it possibly the oldest living thing on Earth.
Do this	• Look for the tallest, biggest tree in the center of a group; it's probably the "mother." • In spring, feel the exceptionally soft new leaves.

For your sketches, notes, observations

	Asters
	Aster species
About the names	• <u>Aster</u> means "star," referring to the look of the flowers.
Memory aid	• Beautiful star-burst flowers: a st[a]r.
When in bloom	• August-October.
Interesting growth habits	• Over 600 species world-wide, about 150 in North America, about 75 in the East—many of them difficult to identify. • Asters grow in every habitat, from wet to dry and shady to sunny. • When the Asters begin to bloom, winter can't be too far behind! • It has been reported that the disk flowers of White Wood Aster, or of other late-blooming white asters, turn reddish only after they have been pollinated. • Because each tiny disk flower has its own pollen, each visiting insect can carry pollen from hundreds of flowers to hundreds of other flowers on another plant.
Warnings, uses	• Native Americans used the showy New England Aster to make a tea for diarrhea and fever. • The Large-Leaved Aster's youngest leaves can be cooked like spinach. • Many butterflies feed on the New England Aster. • Several of the Asters were thought by Native Americans to bring success in hunting.
"Stories"	• The Michaelmas Daisies of England? They are derived from New England and New York Asters, sent back to England by the settlers (Michaelmas, September 29, honors Archangel Michael). • A Greek legend says that Aster s were made from star-dust.
Do this	• Aster species differ in some noticeable and some not so noticeable ways. Examine several kinds and compare such things as: leaves (are they hairy or not? clasping the stem or not? stiff or soft? long or heart-shaped or something else?); stems (smooth or not? zig-zag? branching?); flowers (what color? how many rays? how many disk flowers? arrangement of bracts under them?). How tall is the plant? What habitat? Et cetera!!

For your sketches, notes, observations

| | **Avenses** |
	Geum species
About the names	• <u>Avens</u> is a name from the Old French, derivation not known. • <u>Geum</u> was the name given to the genus by Pliny (23-79 A.D., he who died in the eruption of Mt. Vesuvius), from a word meaning "pleasantly fragrant," referring to the smell of the freshly-dug root.
Memory aid	• Flower like a pale yellow, cream or white and green star: star of [he]avens.
When in bloom	• June-August; pods August-September.
Interesting growth habits	• Five green pointy sepals are often longer tha n five petals and stick out from under them. • Seed head is a bristly cluster of hooked seeds that hang down and all around in a ball, catching on clothing. • Leaves in threes with center one biggest. • Makes low evergreen rosette in winter.
Warnings, uses	• Actually, you may be used by the Avens to transport its seeds. • We can't resist noting that the roots of one Avens, Water or Purple Avens, are described as making a "chocolate-like beverage...the flavor is much improved with the addition of milk and sugar." Probably a bit of Swiss Miss would help too.
"Stories"	• Avens is one of those somewhat nondescript, weedy flowers that get overlooked in favor of showier plants. But a close look is rewarding. • "...a neighbor of mine in hay-time, having overheat himself, and melted his grease, with striving to outmowe another man, fell dangerously sick, not being able to turn himself in his bed, his stomach gon, and his heart fainting...I administered...Avens roots and leaves in water and wine, sweetening it with syrup of Clove-Gilliflowers; in one week's time it recovered him." (1674)
Do this	• If seed heads are present, use a magnifier to admire the ingenious seeds.

For your sketches, notes, observations

	Clammy Azalea, Swamp Azalea, Swamp Honeysuckle *Rhododendron viscosum*
About the names	• <u>Azalea</u> comes from Greek, meaning "from dry habitats," appropriate for many of this genus though not this one, which instead grows in wetlands (hence "<u>swamp</u>"); <u>clammy</u> means "soft, moist, and sticky; viscous," referring to the flowers; <u>honeysuckle</u> is appropriate because of the sweet odor. • <u>Rhododendron</u> means "rose tree", first used in 1601 to describe this genus, on account of its showy flowers; <u>viscosum</u> means "viscous," describing the sticky buds and flowers.
Memory aid	• The scent of the flowers is so strong it will assail - ya [azalea] as you walk by the swamp.
When in bloom	• June-July, but as late as September.
Interesting growth habits	• Leaves form a kind of spray around the bud. • Long, curved stamens stick way out of the flower, past the petals, which are joined to form a tube. • Blooms late for an Azalea, after other species. • Not evergreen.
Warnings, uses	• Not of much interest to large wildlife, but certainly its nectar is valuable to pollinators. • A joy to find and smell while walking!
"Stories"	• Seeds of this plant were sent back to England in 1680 by the Virginia Colonists.
Do this	• SMELL THE FLOWERS! • Feel the sticky flowers; use a magnifier to look for the tiny reddish hairs which are the cause of the stickiness. • Look to see how stamens outline an oval shape, directing a pollinator to the landing area, and how the female pistil is longer than the stamens, so it is touched first, insuring that the pollinator drops off pollen from the previously-visited flower before getting its nectar reward within.

For your sketches, notes, observations

| | **Japanese Barberry** |
	Berberis thunbergii
About the names	• <u>Barberry</u> and <u>Berberis:</u> both from an Arabic name for the peoples of North Africa; <u>Japanese:</u> distribution maps show the family in Japan, source of this species, and in North Africa. • <u>Thunbergii:</u> for K. P. Thunberg (1743-1822) of Uppsala, Sweden.
Memory aid	• Think: <u>barb</u> <u>berry</u>. Japanese Barberry has a single thorn and single berry (compare to Common Barberry, which has a triple thorn and clusters of berries).
When in bloom	• April-May; fruits later, persistent in winter.
Interesting growth habits	• An invasive alien! • Among first shrubs to leaf out in spring, thus getting a head start and shading out other plants. • Other barberries are highly subject to a fungus which requires grains as one of its two hosts; Japanese Barberry is resistant to this fungus. • Flower has complex trigger mechanism in stamens for shooting pollen onto a visiting bee.
Warnings, uses	• Avoid letting this invasive into your garden! • Fruits eaten by many birds, including pheasant, bobwhite, ruffed grouse—thus spreading the plant around the forest.
"Stories"	• Brought to New England as seed around 1875, to be used as an ornamental. • A group of New England Wild Flower Society volunteers, searching for an endangered species in the Berkshires, reported, "What lay before us, as far as the eye could see, was...Japanese Barberry. We had to force and hack our way through more than half a mile of these thorny shrubs. Under their dense cover, virtually nothing but Barberry seedlings was able to grow."
Do this	• Dig it out of the woods and burn it! • If it is spring, test the pollen-shooting device by touching a pin or pine needle to the base of one of the stamens. It will reset itself in a few minutes.

For your sketches, notes, observations

	American Beech *Fagus grandifolia*
About the names	• <u>American Beech</u>: comes from the same ancient Germanic root as the word "book," since its smooth bark invited writing and was used for the same (there are also European Beeches). • <u>Fagus</u>: Latin name for this tree; <u>grandifolia</u> means "large-leaved."
Memory aid	• Bark is smoothly granular like sand on the beach; leaves have waves like waves on the beach.
When in bloom	• April-May; fruit (nuts) Sept.-October.
Interesting growth habits	• Among the most shade-tolerant of our trees, they can sprout in their own shade and under other trees. • Pale dry leaves, like cream-colored tissue paper, flutter on young Beeches in winter. • A small grove usually represents one parent tree and numerous smaller cloned trees, arising from underground suckers. • Beeches can live three to four hundred years. • All Beeches in a range produce an exceptionally large crop of nuts every two to three years; this is called masting, and is probably a device to prevent nut-eaters from consuming all the trees' nuts in any one year—if there are so many at once, the eaters can't hope to eat them all (a tactic similar to simultaneous births in herds of some animal species).
Warnings, uses	• Nuts are edible to people and other animals, such as porcupines, deer, flying squirrels, turkeys, possums, grouse, bears, raccoons —a veritable manna! • Colonists used dry leaves to stuff mattresses. • A wash of one ounce tea to one pint salt water was used for burns, frostbite, and poison ivy.
"Stories"	• In bear country, you might be rewarded by spying sets of parallel scratches on the Beech bark, made by a bear climbing up to get nuts.
Do this	• Look at the ends of twigs for the sleek, sharp, elegant buds. • Check nearby for Beechdrops, parasitic on Beech roots: low cluster of thin brown stems with a purplish cast, blooming modestly August-October.

For your sketches, notes, observations

CAUTION	**Black Birch**
	Betula lenta
About the names	• <u>Black</u> refers to dark gray or mahogany-colored bark; <u>Birch</u> from an ancient word meaning "to shine, white" ("bright" is related). • <u>Betula</u>: name given by Pliny in ancient times, meaning "pitch," since bitumen is distilled from bark; <u>lenta</u> means "pliable," probably referring to twigs. • Also called Sweet Birch, Cherry Birch.
Memory aid	• Scratch and sniff test the best, just as for Cherry: when in doubt, sniff. If it smells <u>B</u>eautifully of wintergreen, and <u>B</u>ark is <u>B</u>lack, it's <u>B</u>lack <u>B</u>irch.
When in bloom	• April-May; fruits August-October.
Interesting growth habits	• Looks a lot like a Black Cherry tree, with its reddish dark bark and horizontal lenticels for gas exchange. • Bark on big older trees separates into scaly plates.
Warnings, uses	• CAUTION Oil is toxic and is absorbed through skin. • A good wood for fine furniture. • When tapped like a Maple tree, the fermented sap can be made into birch beer. • Many parts eaten by forest creatures. • Native Americans used tea for fever, stomach, lung, bladder problems; and as pain-reliever.
"Stories"	• Can't sprout except on bare ground; also fire-sensitive, so a forest with many large Black Birches suggests disturbance from logging or storm, but not from fire. • During 19th c. in Appalachia, Black Birch's wintergreen oil was harvested in great amounts, resulting in destruction of entire forests (later made artificially, thus sparing the Birches).
Do this	• Scratch bark down to the green and sniff for its wintergreen scent. • Turn leaf to see tufts of white hair on veins. • Look in spring for drooping male and upright female catkins. • In winter, look on snow for tiny winged "angel" seeds.

For your sketches, notes, observations

	Gray Birch *Betula populifolia*
About the names	• <u>Birch</u>: from an ancient word meaning "to shine, white" (a related word is "bright"); <u>Gray</u> for the grayish-white bark. • <u>Betula</u> is the Latin name for "birch," (given by Pliny before 79 A.D.); bitumen, which is made from the bark, comes from the same r oot word; <u>populifolia</u> means "poplar-leaved," referring to the leaf shape; name derived from Latin words meaning "tree of the people." • May also be called Wire Birch or (mistakenly) White Birch.
Memory aid	• Leaves are triangular and very long -pointed, resembling a "Y" as in "graY"; bark shows black chevrons (^), which somewhat resemble upside - down Y.
When in bloom	• April-May; fruits in September.
Interesting growth habits	• Unlike bark of White Birch, Gray Birch bark does not peel off tree. • Lots of black chevrons and black horizontal lines on trunk. • Pioneer species on burned or cleared areas; because it grows quickly and doesn't live long, it's sometimes considered a "weed" tree.
Warnings, uses	• Used to make clothespins and spools; twigs can be bound together to make rough brooms. • Birds enjoy the seeds (found in tiny cone -like structures).
"Stories"	• In a heavy snow, Gray Birch branches as tall as 30 feet or more may bend all the way to the ground without breaking.
Do this	• If there is snow on the ground, look under the Gray Birch for its seeds that look like tiny angels. • In spring, look for male catkin (usually just one) hanging from tips of branches, and female catkins that stand up straight, away from tips.

For your sketches, notes, observations

| | **Paper/White Birch** |
	Betula papyrifera
About the names	• Paper/White: bark is brilliant white and peels off like paper; Birch comes from Old English word meaning "bright," "white," or "to shine." • Betula: name given by Pliny in ancient times, meaning "pitch," since bitumen is distilled from the bark; papyrifera: "paper-bearing." • Also called Canoe or Silver Birch.
Memory aid	• Unmistakable; note that whitish bark of Gray Birch does not peel as does that of this tree.
When in bloom	• April-May; fruits August-September.
Interesting growth habits	• Among the first trees to grow in an area that's been burned or cut over; lives only 60-80 years. • Bark has 6-9 layers; color may prevent tree from overheating by reflecting sunlight. • Male catkin flowers in groups of 2-3 hang off twig tips when flowering (erect in winter); female flowers grow upright in leaf axils. • In bloom, male catkins are three inches long; each makes over five million pollen grains.
Warnings, uses	• Wood used for ice cream bar sticks, toothpicks. • Native Americans used the bark for numerous items: canoes, boxes, cups, house coverings.
"Stories"	• Leaves have high sugar content, so earthworms enjoy them in leaflitter on forest floor. • Tendrils of bark will ignite even when damp, useful when you're lost and cold in the woods! • To make a canoe, stretch big sheets of bark over a framework of cedar and sew them on with spruce roots; seal it with spruce pitch...how many arts we have forgotten.
Do this	• Admire it! Especially the delightful rosy color of the inner bark, or a grove of them in the snow, against a deep blue sky. • Feel the chalky texture of the bark. • Look for horizontal marks called lenticels, openings into the bark for exchange of gasses. • In winter, look for tiny "angels" or "stars" on the snow near White Birch; these are the seeds, fallen from the catkin-cones. • One thing NOT to do is to remove bark from a living tree.

For your sketches, notes, observations

	Yellow Birch *Betula lutea*
About the names	• <u>Yellow</u>: refers to golden quality of bark; <u>Birch</u> is from Old English word meaning "bright, white," or "to shine." • <u>Betula</u>: name given by Pliny in ancient times, meaning "pitch," since bitumen is distilled from bark; <u>lutea</u> means "yellow."
Memory aid	• Glowing golden peeling, curling bark, like the yellow feathers of a golden birch-bird.
When in bloom	• April-May; fruits August-October.
Interesting growth habits	• Most shade-tolerant Birch. • Bark is highly flammable. • Has to have clear ground for sprouting, so probably germinates after fires. • Grows fast, perhaps 10 feet in six years. • Tiny flowers: male are yellowish on branch tips in drooping catkins, female are greenish in upright catkins behind tips. • Fruits look like small upright cones, and a stand of Yellow Birch may make over a million seeds per acre.
Warnings, uses	• Sap can be boiled to make a sweet syrup; twigs steeped in hot water for tea; used to make birch beer. • Lumber used for fine furniture.
"Stories"	• Tree may germinate on top of a rotting log, and send roots into the ground below; when the log rots away, the curving roots are left, like big curving stilts on which the tree rests.
Do this	• Scratch and sniff a twig for the wintergreen scent; also try sniffing a yellow fall leaf. • Look for a tree up on stilts. • Once identified, the glowing golden and lacy quality of this tree provides a sweet surprise in the forest.

For your sketches, notes, observations

POISONOUS	**Oriental Bittersweet**
	Celastrus orbiculatus
About the names	• <u>Oriental</u>: introduced from Asia; <u>Bittersweet</u>: a confusion with another plant named the same, whose fruits very slightly resemble these.
	• <u>Celastrus</u>: name given to an evergreen tree by Theophrastus (372-287 B.C., Aristotle's successor, who wrote extensively on botany); <u>orbiculatus</u>: circular, referring to the leaves.
Memory aid	• It's so destructive and so pretty: bitter-sweet.
When in bloom	• May-June; fruits October-April.
Interesting growth habits	• Vines spread rapidly and aggressively, curling around and over shrubs and trees, often shading them out or suffocating and pulling them down.
	• If there's nothing to climb on, plant just rummages around on the ground until it can find a place to go up, as much as sixty feet; stems will twine around each other and just stand up and reach out.
	• Some plants are male (no fruit), others female (bear fruit).
Warnings, uses	• CAUTION Contains a poisonous compound; NOTE that holiday arrangements of the vines and fruit are accessible to and may be attractive to children.
	• Also, of course, watch your ankles when walking near it; you don't want to be grabbed!
	• The vines, with their pretty yellow seed capsules and red seeds within, are often used to make holiday wreaths; then, when the holiday is over, they are tossed away, to sprout anew on the trash pile.
	• Seeds are eaten and passed along by many birds.
"Stories"	• One source says, hey, if you want an endless supply for decorations, just plant some in your garden. Do Not Do This!
	• Introduced as an ornamental in the late 1800's.
	• Note that there is also an American (non-invasive) Bittersweet, in which the flowers and fruit grow in terminal clusters rather than in axils.
Do this	• In mid-summer, look on stems for black and yellow treehoppers; in fall, look for rounded mounds of hardened froth; within are treehopper eggs.

For your sketches, notes, observations

POISONOUS	**Bittersweet Nightshade**
	Solanum dulcamara
About the names	• <u>Bittersweet</u>: supposedly some parts of the plant taste bitter at first, then very sweet—but don't try this at home!; <u>Nightshade</u>: possibly refers to the dire effects of ingesting this plant. • <u>Solanum</u> means "comforter," "quietening," referring to sedative properties; <u>dulcamara</u> means "bittersweet."
Memory aid	• Plant is poisonous (bitter) and flower is beautiful (sweet); purple flower petals are curled over like a shade against the yellow star in the center(night-shade).
When in bloom	• May-September.
Interesting growth habits	• May have flowers, green berries, and ripe red berries on one plant at the same time. • Although yellow center of flower looks like pollen, in fact pollen is hidden inside, and there's no nectar. Insects go for the yellow, buzz around looking for nectar; the buzzing vibrates pollen loose onto their bodies. • Does not climb by tendrils or twining, but just sprawls and flops on whatever's handy. • Leaves always face the sun and flower clusters always face in a direction different from the leaves.
Warnings, uses	• CAUTION All parts of the plant are poisonous, even when dried, due to an alkaloid compound. • Many birds and mammals eat the berries. • Found to have anti-cancer properties. • Used to make steroids.
"Stories"	• In the tomato family, related to potatoes, eggplant, tobacco. • Pliny (died 79 A. D.) recommended its juice "for those who have fallen from high places" (does this mean literally or figuratively, or both?). • Wearing a garland of this plant will keep witches and witchcraft at bay.
Do this	• Use a magnifier to admire the exotic flower with its backswept purple petals; look for the hidden pollen. • Check to see if flowers face away from leaves.

For your sketches, notes, observations

| | **Common Bladderwort** |
	Utricularia vulgaris
About the names	• <u>Bladder</u>: for the tiny bladders on the roots; <u>wort</u> means "plant." • <u>Utricularia</u> means "little bladder or "little bottle"; <u>vulgaris</u> means "common."
Memory aid	• Combination of <u>Bladder</u>wort plus <u>yellow</u> flowers plus being a <u>water</u>-growing plant might suggest a way to remember the name.
When in bloom	• May-August.
Interesting growth habits	• Plant not attached to floor of pond in which it grows; floats free with a mass of hair-like leaves below yellow flower, which rises on a straight stem several inches above surface of water. • Lures and captures zooplankton! as follows: • Up to 600 minute bladders on plant's submerged leaves. • Each bladder equipped with trapdoor baited with sugars. • Bladders are deflated; when touched by an organism, suddenly inflate with water and suck creature in, closing in 2/1000ths of a second. • Organism digested by plant enzymes in 15 minutes to two hours. • Nutrient-filled water now in bladder absorbed by plant, creating new vacuum and resetting trap. • Organisms have been seen half in and half out of bladder, and half-digested. • Isn't that a fabulous story!
Warnings, uses	• Have been known to eat small pets (Ed. note: She is probably joking.)
"Stories"	• Charles Darwin first established the purpose of the bladders, in 1875; before then they were thought to be flotation devices.
Do this	• Plant may be lifted from water without damage to it to examine its tiny bladders.

For your sketches, notes, observations

POISONOUS	**Bluebead Lily/Clintonia**
	Clintonia borealis

About the names	• <u>Bluebead</u>: fruits of this member of the <u>Lily</u> family are round and bright blue; often called by its genus name. • <u>Clintonia</u>: for DeWitt Clinton (1769-1828), 3-term mayor of New York City, governor of New York, sponsor of the Erie Canal, and naturalist; <u>borealis</u> means "of the north." • Also called Corn-Lily, Bear-Plum, Dragoness-Plant, Bear's-Corn, Yellow Clintonia, and many others.
Memory aid	• Easy when in fruit; just describe the berries. • Usually 3 broad, smooth, shining bright green basal leaves, slightly leathery, from which rises the bare flower stalk, bearing drooping greenish-yellow bell-like flowers: use your own view of a past president to concoct a memory aid!
When in bloom	• May-June/July; fruits follow.
Interesting growth habits	• Like many another woodland plant, Clintonia cannot be transplanted, since it requires the precise qualities of the mini-habitat in which it is found.
Warnings, uses	• CAUTION: Sources differ as to whether berries are poisonous. Therefore, do not experiment! • Young leaves said to taste of cucumber, but not good after they are unfurled. • Native Americans used leaves on rabid-dog bites, drank leaf tea for heart ailments and diabetes, and used the root to aid in childbirth. • Root contains an anti-inflammatory compound as well as a substance from which progesterone is made. "Science should investigate," says the Peterson's medicinal plants guide.
"Stories"	• One writer tells of meeting, on a forest walk, a person carrying a hatful of "blueberries," gathered from convenient stalks sticking up from the forest floor, doubtless all excited about this great find and ready to turn it into a pie. A rude surprise awaited.
Do this	• Enjoy feeling the smooth, leathery leaves.

For your sketches, notes, observations

	Blueberries
	Vaccinium species
About the names	• <u>Blueberry</u>: there are lots of blue berries, but few if any with that true-blueberry color, tiny star-shaped cup on top side, and whitish bloom. • <u>Vaccinium</u>: "a name of great antiquity with no clear meaning."
Memory aid	• Scraggly zig-zag bunches of warty twigs: Yum?!
When in bloom	• Lowbush species: April-June; fruit June-September ("early"), August-September ("late"). • Highbush species: May-June; fruit June-September.
Interesting growth habits	• Twigs are green in summer; red above and green below in winter. • "Mummy Berries": a fungus grows under plant in spring; this releases spores which infect the leaves, producing a sweet taste and ultraviolet reflection attracting bees; bees take spores from leaves to flowers, which then mature into grayish, hard "berries"; cycle begins again. • Depends on symbiotic relationship with fungi.
Warnings, uses	• Native Americans prescribed the fumes from burning dried flowers for "madness." • Many creatures relish the berries and twigs. • The cultivated berries come from Highbush stock.
"Stories"	• The Bowl-and-Doily spider builds a small web bowl in some twigs, with a bit of webbing above this, and below it, a flat platform (the doily); insects hit the web above the bowl, fall into it and ZOT! the spider gets them! • Bees occasionally "steal" nectar, without gathering pollen, by chewing a hole at the base of the flower and sucking the sweetness out.
Do this	• ONLY IF you are certain of your identification, feast on the berries (leave some for bears, birds, deer, rabbits, other people, and new plants). • In spring, examine the dainty flowers with a magnifier; sniff them for the also-dainty scent. • Open a berry and see if you can find any of the hundreds of minute seeds. • Look for evidence of the Bowl-and-Doily spider.

For your sketches, notes, observations

	Bluets *Houstonia caerulea*
About the names	• <u>Bluet</u> means " little blue," which they are. • <u>Houstonia</u>: for William Houston, 1695-1753, a botanist and physician, traveled as a ship's surgeon to Central America, where he collected plants; <u>caerulea</u> means "sky-blue." • Also called Angel-Eyes, Blue-Eyed Babies, Quaker-Ladies, Innocence, Starlights, Sky-Flower, Eyebright—and many others.
Memory aid	• Bluets have four tiny petals: four letters in the word "blue," and the plant is tiny (2"-7"), thus "bluet," little blue.
When in bloom	• April-June/July.
Interesting growth habits	• Flowers close each night and bend over. • May grow in clusters; they spread by underground runners that come to the surface and grow into new rosettes of leaves. • Two forms of the flower: Type A has tall stigma (female) and short anthers (male) below it; Type B has tall male anthers and short female stigma below it—this enables cross-pollination. • Flowers are sometimes white.
Warnings, uses	• Just enjoy finding it!
"Stories"	• May be pollinated by a fly that looks like a bee, called a bee-fly, less than half an inch long, with partly transparent wings. • In same family as coffee, gardenias, quinine.
Do this	• Try to find two forms of flowers—you have to look closely!

For your sketches, notes, observations

	Bunchberry
	Cornus canadensis
About the names	• <u>Bunchberry</u>: the plant's berries grow in a tight cluster or bunch. • <u>Cornus</u> means "horn" (see story below) and is the name given to the Dogwood genus, of which Bunchberry is a member; <u>canadensis</u> means "of Canada" (since it is either native to Canada or was first described there). • Also called Dwarf Cornel, Bunchplum, Pudding-Berry, Herb Dogwood.
Memory aid	• In the center find a <u>bunch</u> of tiny yellowish flowers; later in the season find a <u>bunch</u> of berries.
When in bloom	• May-July; fruits July-September/October.
Interesting growth habits	• White bracts look like flower petals, but the tiny real flowers are centered between these. • "Flower" looks almost exactly like that of a Dogwood tree! • Usual whorl of six green leaves around the flower may be reduced to four leaves if there is no flower; this may be because producing a flower requires the additional energy to be gotten from the two additional leaves. • May form colonies. • Indicates an acid soil.
Warnings, uses	• Berries are edible raw or cooked, although the taste is described as "insipid." • Enjoy its elegant simplicity!
"Stories"	• All but one other member of this family are trees or shrubs. • The Latin name <u>cornus</u>, meaning "horn," apparently is an allusion to the hardness of Dogwood wood; Virgil (70-13 B.C.) wrote in <u>Aeneid</u>, "They wear on their hair ceremonial garlands, well-trimmed,/And each of them carries a couple of steel-tipped cornel-wood lances."
Do this	• Check to see if plants without flowers have only four leaves surrounding the white bracts and cluster of flowers within, rather than six leaves as on flowered plants.

For your sketches, notes, observations

	Burdock *Arctium minus*
About the ***names***	• <u>Burdock</u>: derived from the Danish word for a bur. • <u>Arctium</u> means "bear" (from the Greek <u>arktos</u>), a name given by Pliny for the shaggy appearance of the burs; <u>minus</u>: "smaller" (as compared to the larger Great Burdock). • Also called Cuckoo-Button.
Memory ***aid***	• The big, heavy, coarse leaves, as well as the burs, would make this a BURD-en to pick up.
When in ***bloom***	• July-October.
Interesting ***growth*** ***habits***	• Overwinters as a low green rosette. • Flower's bracts end with curved hooks, the part that sticks to passersby; seeds are in sheaths attached to the hooks; when hooks are pulled in an attempt to dislodge, sheaths open and release seeds. • Note that Thistles have neither the large leaves nor hooked bracts of Burdock.
Warnings, ***uses***	• Unless you want to bring this plant home, check clothes, and dog, for burs after visiting it. • Flower stalks, after cooking, can be simmered in sugar water to make candy. • "The peeled stalk...boiled in the broth of fat meat, is pleasant to be eaten; being taken in that manner it increaseth seed and stirreth up lust." (1633) • Most parts of young plants can be cooked and eaten in various ways. • Used in traditional Asian medicine. • Used to treat dandruff and scalp infections (probably after removing the burs).
"Stories"	• Studies have suggested the plant contains a compound that prevents mutations. • In <u>As You Like It</u>, Rosalind says to Celia: "How full of briers is this working-day world!" Celia replies: "They are but burs, cousin, thrown upon thee in holiday foolery. If we walk not in the trodden paths, our very petticoats will catch them."
Do this	• Use a magnifier to admire the tiny curved hooks on the bur—Nature's Velcro.

For your sketches, notes, observations

	Butter and Eggs *Linaria vulgaris*
About the names	• <u>Butter and Eggs</u> refers to the yellow and orange colors. • <u>Linaria</u> means "flax-like," since some of this genus resemble Linaceae, the Flax family; <u>vulgaris</u> means "common." • Also called Toadflax, Bread-and-Butter, Eggs-and-Bacon.
Memory aid	• Few other flowers are so distinctly yellow in one part and orange in another: the butter on your egg yolk (buttered toast and a fried egg).
When in bloom	• June-October.
Interesting growth habits	• Looks like a tiny Snapdragon (in the same family); only insects heavy enough to push open the flower can get its nectar and transfer its pollen. • "Neither the spade, plough, nor hoe can eradicate it when it is spread in pasture. Every little fibre that is left will soon increase prodigiously; nay, some people have rolled great heaps of logs upon it, and burnt them to ashes, whereby the earth was burnt half a foot deep, yet it put up again as fresh as ever, covering the ground so close as not to let any grass grow amongst it; and the cattle can't abide it." (1758, John Bartram, by 1765 official botanist to the King of England, and botanist extraordinaire in the American South)
Warnings, uses	• Make a tea with milk as an insecticide.
"Stories"	• Found on early "garden lists" of the settlers, who brought it over to plant in their dooryards to remind them of home. • Orange section of lower lip is a "honey guide" to nectar, visible and attractive to insects. In an experiment, these flowers were placed between sheets of glass; hawk moths homed in on the orange honey guide and pressed their tiny tongues to its location under the glass, leaving the glass marked; when the orange section was removed and stuck on another, different kind of flower, they went for it there too.
Do this	• Gently make this flower "snap"; use a magnifier to look within for the furry pathway leading straight to nectar and waiting pollen.

For your sketches, notes, observations

POISONOUS	**Buttercups** *Ranunculus* species
About the names	• <u>Buttercup</u> describes the shining deep yellow curving petals very evocatively. • <u>Ranunculus</u> means "little frog," because many in the genus live near water.
Memory aid	• The shiny yellow petals look as if they've been greased with butter.
When in bloom	• May-September.
Interesting growth habits	• The shiny look of the petals is due to a layer of special starchy cells just under their surface. • At the base of each petal is a scale with a tiny nectar gland. • Usually have many stamens and pistils; looks bushy in flower center. • Overwinters as a low green rosette.
Warnings, uses	• CAUTION Contains a toxic compound that can cause blistering and severe gastric problems if handled or eaten; in large quantities has been known to cause paralysis of the nervous system.
"Stories"	• To make your cows give creamy milk, rub their udders with Buttercups, or hang a bunch of them over the barn door (the flowers, not the cows, who will be mooing in pain from their blisters). • One reference says that beggars used these plants, with their blistering quality, to keep open sores on their bodies. • Brought to America by early settlers as a garden flower.
Do this	• Enjoy looking at the crowded center of the flower with a magnifier; look at petal base for scale and nectar glands. • With a fingernail, gently scrape the yellow petal: surprise! you can see through it!

For your sketches, notes, observations

	White Campion or Evening Lychnis
	Lychnis alba
About the names	• Origin of word <u>Campion</u> is not known; <u>Lychnis</u> means "lamp" because the hairy leaves were used as wicks in oil lamps; <u>Evening</u> refers to the fact that the flowers open at night; <u>alba</u> means "white."
Memory aid	• Champion Campion, the plant with the biggest bladder! (Calyx, just below flower, is inflated into large bladder shape.)
When in bloom	• May-September.
Interesting growth habits	• Female flowers, with 5 curved styles sticking out of center, have inflated and sticky calyx with 20 veins on it; male flowers, on another plant, have 10 stamens and a slender calyx with 10 veins. • Flowers open at night and therefore are pollinated by moths. • In winter, the seed container is like a shiny tan vase, with 8-10 pointed teeth around upper edge. If the wind hasn't blown it around too much, the little vase will be filled with tiny seeds.
Warnings, uses	• May invade grain and other crops, including nurseries, as a weed.
"Stories"	• Distinguishing the various flowers in these groups is acknowledged to be difficult and controversial: "they" haven't decided everything! • Referring to the fact that some plants are male and some female, Erasmus Darwin (Charles' grandfather) wrote: "When the females arrive at their maturity, they rise above the petals, as if looking abroad for their distant husbands." (1789) • If you pick this plant, either (a) your mother will die or (b) you will be struck by lightning.
Do this	• Look around to see if you can find both male and female plants. Count the veins on the calyx to decide. • If seeds are present, use a magnifier to see their amazing details!

For your sketches, notes, observations

	Cattails
	Typha species
About the names	• <u>Cattail</u>: long, thin, brown and fuzzy.
	• <u>Typha</u>: the Greek name for this genus.
Memory aid	• A cat's brown furry tail would definitely stick straight up if he found himself in the water!
When in bloom	• May-July.
Interesting growth habits	• Flower spike has male and female sections: on the upper part are male flowers, which shed pollen into female flowers (dark brown) below them.
	• One seed head can contain up to 220,000 seeds.
	• Dark brown seeds are each attached to stem by a tiny stalk, which has many tiny hairs on it; all is packed tightly into the spike, and when seeds are ripe, the spike puffs open in fluffy clumps and scatters seeds on the wind.
	• One seed in one season may yield a rhizome clump 10 feet wide with 100 new shoots.
Warnings, uses	• Uses worldwide are legion: fans, burial shrouds, sandals, chair seats, toys, barrel caulking, thatch, cushioning for horses' necks, spear handles, rope, paper, bedding, socks and slippers, quilts, diapers, torches, mattresses, baseball filling, life-jackets—and people eat all parts, from the heart to the pollen.
	• Flour can be made from the pollen (mixed with wheat flour) and from the roots, through an elaborate process of crushing and washing.
	• Young flower spikes can be boiled and eaten like corn on the cob.
	• Native Americans used many parts of the plant for all sorts of medicinal purposes, including treatment for burns, diaper rash, and diarrhea.
	• Muskrats use stalks for their lodges and all parts for food; birds use the plants for cover; mice and birds use the seed fluff for nesting.
"Stories"	• Cattail facilitates marsh development as roots trap decaying material which becomes soil.
	• A spider folds over leaf tips, lines the enclosure with silk, lays eggs within and dies, leaving her body to nourish hatching young.
Do this	• Look among the Cattails and listen for the tiny marsh wren with its reedy, gurgling song and for the redwing blackbird (conk-a-ree).

For your sketches, notes, observations

POISONOUS	**Celandine**
	Chelidonium majus
About the names	• Celandine/Chelidonium means "of the swallow" (the bird) because the plant was supposed to arrive with the swallows and fade when they depart; majus means "larger." • Also called Devils'-Milk, Sightwort, Swallowwort, Wartflower, Wartwort (how do you say this??).
Memory aid	• Flowers and sap are deep yellow: think "yella - cela."
When in bloom	• April-August/September.
Interesting growth habits	• Sap from any part of the plant stains deep yellow.
Warnings, uses	• CAUTION: Sap may irritate skin. • CAUTION: Contains many alkaloids, which in overdose can cause digestive and circulatory distress. • Sap used to cure warts, ringworm, corns, and eczema, and to remove freckles. • Because the sap is yellow, like bile (secreted by the liver), it was used in liver disorders.
"Stories"	• "Some have related that if any of the swallows' young ones be blind, the dams bringing this herb, do heal the blindness." (1st c. A. D.) • "The juice is good to sharpen the sight, for it...consumeth awaie slimie things that cleave about the ball of the eye." (1633) • "Ask now the fouls of the air and they shall tell thee. The swallows will carry thee to the Celandine. Feeble Eyes will not find a greater friend in the whole of the vegetable kingdom." (1724) • An alien (originally from Eurasia), brought by early settlers for its medicinal uses.
Do this	• Carefully break a stem and squeeze to see the beautiful sap; avoid getting juice on your skin in case you are sensitive to it.

For your sketches, notes, observations

| POISONOUS | **Cherries** |
	Prunus species
About the names	• <u>Cherry</u>: a very ancient Semitic word for this plant (cf. "cerise" the color). • <u>Prunus</u>: Latin name for plum tree. • Black, Pin, and Choke Cherry in our area, differing in size and a few other particulars.
Memory aid	• Scratch and sniff the bark: disagreeable odor smells like stale cigars; visualize person sitting in a chair (CHERR-y) and puffing away.
When in bloom	• April-July; fruits July-October.
Interesting growth habits	• Often find Black Knot Fungus growths on branches (another way of identifying Cherry). • Grows in burned areas as pioneer species. • Bark usually has horizontal or dot-like lenticels, openings for gas exchange.
Warnings, uses	• CAUTION That is a cyanide compound that gives the nasty smell; all parts are poisonous (except fruit without the seed), and children have been poisoned from chewing the twigs. • The fruit, while edible, is extremely sour (hence "Choke" Cherry); can be used for jelly. • Black Cherry wood used for fine furniture. • Over 50 species of birds and most forest mammals feed on fruits or twigs. • Accumulates important nutrients as it grows, thus enriching and restoring soil. • Syrup used to flavor cough medicines.
"Stories"	• 1634: "[the fruits] so furre the mouth that the tongue will cleave to the roofe, and the throat wax horse with swallowing those red Bullies." • Sites in New Hampshire showed over 150,000 Pin Cherry seeds per acre. • Passing through a bird's gut improves the germinating ability of the seed.
Do this	• Do scratch and sniff a twig to fix the smell in mind. • Look for two tiny black glands on the leaf stalk. • Look near the ends of twigs for a small shiny cluster of brown "bubbles"; this is the egg mass of the tent caterpillar; look also for their web nests and the caterpillars within, in spring.

For your sketches, notes, observations

	Rough-Fruited Cinquefoil
	Potentilla recta
About the names	• <u>Rough-Fruited</u> refers to the hairy seed pods; <u>Cinquefoil</u> is French for "five-leaf" (from ancient Greek meaning the same thing). • <u>Potentilla</u>: "little powerful one" for its medicinal powers; <u>recta</u> means "upright." • Also called Five-Fingers.
Memory aid	• Five petals, five sepals, five leaflets, so if you can remember the French "cinque" (5), you are all set.
When in bloom	• May-August.
Interesting growth habits	• There are a number of different Cinquefoil species, including a common creeping form that looks like a cross between a buttercup and a strawberry. • Fall remains (actually just dried flower bracts) look like tiny brown rosebuds.
Warnings, uses	• Used as an astringent to stop nosebleeds, and as a remedy for fever and toothache.
"Stories"	• The five leaflets were thought to symbolize the five senses, and only knights of yore who had achieved self-mastery were allowed to use the leaf as a device on their shields. • Add a Cinquefoil leaf to your fishing net to increase your catch. • Lovers! make a potion of the leaves to lure your desired one. • "This is an herb of [the planet] Jupiter, and therefore strengthens the parts of the body it rules; let Jupiter be angular and strong when it is gather'd, and if you give but a scruple (which is but twenty grains) of it at a time, either in white-wine or whitewine vinegar, you shall very seldom miss the cure of an ague [like malaria]." (1652)
Do this	• Use a magnifier to see the hairy, glandular stems. • In fall and winter, admire the five-pointed little seed cups; one source describes them as tiny rosebuds. • If there are curled-up dry leaves left, unfold one to see the five-parted leaf of summer.

For your sketches, notes, observations

	Clovers
	Trifolium species
About the names	• <u>Clover</u> from an ancient German word for this plant. • <u>Trifolium</u> for its leaves in threes (and, if you're lucky, a set of four now and again).
Memory aid	• These plants recognized by most everyone.
When in bloom	• May-October.
Interesting growth habits	• Legumes (nitrogen-fixers) which, by means of bacteria living in nodules among the roots, capture atmospheric nitrogen and convert it to a form usable by plants. • Seed-coats are very hard, enabling them to remain dormant for a long time. • Flower heads may have as many as 40-60 florets. • Different species pollinated by different bees, depending on size; they have to work to get into the flowers.
Warnings, uses	• Rich in protein; make flour from the flowers and seeds, but "not among the choicest of wild foods." • A substance found in Red Clover is now being studied for its possible effectiveness in fighting diabetes and AIDS.
"Stories"	• More on Legumes: the bacteria need an oxygen-free environment, and hemoglobin traps oxygen. Hemoglobin is made of a protein (supplied by the Legume plant) and the chemical heme (supplied by the bacteria). So the Legumes and the bacteria combine to enable the bacteria to survive and capture nitrogen to nourish the plants. Wow. So never say: "It's just Clover"! • Note that if you have Clover in your lawn, you have natural fertilizer along with it. • Red Clover is Vermont's state flower. • Red Clover was thought in ancient times to protect against witchcraft and evil spirits.
Do this	• If there are lots of Clovers in a patch, gently pull one up and look for the nodules among the roots; they are pink because of the hemoglobin in them. • Use a magnifier to examine the complex flower with its upper and central "standard" petal, its two fused lower petals, the "keel," and the lateral "wings."

For your sketches, notes, observations

CAUTION	**Clubmosses**
	Lycopodium species
About the names	• <u>Clubmoss</u>: first used in 1597 to name one of the species, which has upright club-shaped spore containers.
	• <u>Lycopodium</u>: "club-shaped wolf's claw," for the club-shaped spore cases on some species, and the claw-shaped roots.
	• Five species seen in this area: Shining, Bristly, Ground Cedar, Ground Pine, Staghorn.
Memory aid	• Grow low to the ground and are evergreen, like <u>mosses</u>, but belong to their own special <u>club</u>.
When in fruit	• Not a flowering plant; spores appear in fall.
Interesting growth habits	• Evergreen.
	• Most of the species look like baby trees.
	• Reproduces not by seeds but by spores; also spreads by underground root stocks.
Warnings, uses	• CAUTION Spores may irritate nose and throat.
	• Spores are so fine that they were used as a flash powder in photography, and stage designers ignited them to create "lightning."
	• Spores used as a kind of soothing powder for many irritations (though since they irritate delicate membranes, this can't be recommended!).
	• Because the spores are uniform in size, they were once used in microscopic measurement.
"Stories"	• These are not actually mosses, but are in their own order.
	• They are ancient plants, more than two hundred million years old (even before dinosaurs or insects); some of them grew to one hundred feet! Where they formed thick mats, dead plants upon dead plants, after millennia these became coal.
	• They have a simple vascular system to conduct nutrients, sort of halfway in sophistication between true mosses and the "higher" plants.
	• Along with ferns, first plants to develop roots, stems, and leaves.
	• Spores produce a very tiny plant with sexual parts, from which develops the Clubmoss seen in the woods; it may take twenty years for the entire growth sequence from spore to spore.
Do this	• Tap the spore containers to see a puff of yellow spores.

For your sketches, notes, observations

	Dandelion *Taraxacum officinale*
About the names	• <u>Dandelion</u>: French <u>dents de lion</u>, "lion's teeth," referring to the deeply toothed leaf margins. • <u>Taraxacum</u> means "disturber," from a Persian name for a bitter herb; <u>officinale</u> means "of apothecaries," from its medicinal history.
Memory aid	• One flower known by all! (Be sure it's not a Hawkweed, though, with branching flower stalk.)
When in bloom	• March/May-June; second flowering in September.
Interesting growth habits	• Only one flower per stem, rising from an evergreen basal rosette; flower closes in evening and opens in morning. • Flowers do not require pollination to produce viable seed.
Warnings, uses	• Numerous long-standing food uses for all parts of this plant: salads, cooked greens, wine, fritters, cooked like carrots, coffee substitute — and on and on. • Valuable source of nectar and pollen for bees. • A diuretic (hence its French name Pissenlit —piss in bed!) and for kidney ailments. • Has a compound potent against yeast infections. • Boiling the flowers gives a yellow dye; boiling the roots, a magenta color.
"Stories"	• Grown for food in New England before 1700. • In 1871, four varieties were exhibited at the Massachusetts Horticultural Society, and by 1884 a man named Corey, of Brookline, was providing the Boston market with seed. • Leaves have more Vitamin A than most greens we commonly eat. • "Young dandelion leaves make delicious sandwiches, the tender leaves being laid between slices of bread and butter and sprinkled with salt...a little lemon-juice and pepper varies the flavor. The leaves should always be torn to pieces, rather than cut, in order to keep the flavour."
Do this	• Using a magnifier, examine one of the tiny individual ray flowers, with male stamen and female pistil, and 5-parted corolla of fused petals; or, if plant is in seed, take a look at the beautiful and ingenious "parachute."

For your sketches, notes, observations

CAUTION	**Dock**
	Rumex crispus
About the names	• <u>Dock</u>: from an Old English word for this plant. • <u>Rumex</u>: used by Pliny (died 79 A.D.) for a plant in this genus; <u>crispus</u> means "waved or curled margin," referring to the leaves of this plant. • Also called Out-Sting, Coffee Weed.
Memory aid	• <u>Dock</u> seeds are like heart-shaped <u>lock</u>-ets.
When in bloom	• June-September; fruits by mid-July.
Interesting growth habits	• Evergreen rosette through the winter. • Nondescript flowers are green; this suggests, correctly, that they are wind-pollinated. • Leaves and stalk often reddish or spotted. • Handsome red-brown seeds often stay on the plant all winter. • Yellow taproot may grow up to six feet deep.
Warnings, uses	• CAUTION Too much of this may cause stomach upsets. • Useful as a laxative. • "Nettle out, Dock in,/Dock remove the nettle's sting." • Grind up the root and use for tooth powder; roast the seeds for coffee; cook the leaves like spinach and serve with vinegar; or eat very young leaves raw in salad. • Contains lots of protein and Vitamins A and C. • Effective against ringworm. • Dried sprays of seed pods make fine decorations.
"Stories"	• You can cure elf sickness (laid on one by witches) with a mixture of Dock leaves, ale, holy water, and other herbs. Aren't you glad to know this? • Seeds have protective coatings which keep them safe if eaten by an animal.
Do this	• In fall, by all means examine the beautiful brown seeds under a magnifier. • Look for the grooved stem; does it have red markings?

For your sketches, notes, observations

Dodder
Cuscuta gronovii

About the names	• <u>Dodder</u> may come from an ancient Dutch word meaning "yolk of an egg" which would refer to the deep yellow or orange of this plant. • <u>Cuscuta</u> is the medieval Latin name for this genus; <u>gronovii</u> is for Jan Fredrick Gronov, a Dutch botanist (1690-1762) who studied American plants and was the teacher of Linnaeus. • Also called Hellbind, Lover's Knot, Love Vine.
Memory aid	• Doddering, rich old man can't stand upright, he's so laden with yellow gold.
When in bloom	• July-October.
Interesting growth habits	• Has no chlorophyll, and no leaves, just a few scales. • Parasitic; can completely envelop another plant in a spaghetti-like tangle of its orange stems. • Although it will germinate in soil, once it attaches with suckers to a host plant, it gets all its nourishment from the host through the suckers, and the roots actually die off.
Warnings, uses	• Watch your foot, there it comes! • Used to treat kidney and liver problems; "an infusion acts as a brisk purge."
"Stories"	• Here's how to find if your lover is faithful: pick a piece of Dodder while thinking of him/her, toss it over your shoulder into the plants it grows on, and go away for a day; if, when you come back, the Dodder has reattached itself to its host, you can breathe easy; if it hasn't, well... • "All dodders are under Saturn...it is the most effectual for melancholy diseases...also for the trembling of the heart, faintings, and swoonings...and for melancholy that arises from the windiness of the hypochondria." (1653, Nicholas Culpeper, who incurred the wrath of the medical establishment by translating their knowledge from Latin into English; unfortunately, he was also deep into astrology.)
Do this	• Look closely in leaf axils to see if you can find its clusters of waxy whitish flowers, just 1/8 inch wide. • Stems twine counterclockwise; look for this.

For your sketches, notes, observations

| | **Duckweeds** |
	Lemna species and other genera
About the names	• Duckweed: grows in profusion, is eaten by ducks.
	• Lemna: name given to water-plants by Theophrastus (372-287 B.C., Aristotle's successor, who wrote technical works on plants).
	• Also called Duck-Meat, Watermeal.
Memory aid	• Fills a duck habitat like a weed.
When in bloom	• Plant appears on ponds in spring.
Interesting growth habits	• Has no leaves, stems, or real roots , but instead is just one body, called the thallus, with a tiny rootlet hanging free in water.
	• Nutrients absorbed by entire under-surface.
	• Certain species produce the tiniest flowers in the world: these entire plants are only 0.5 or 0.7 mm. wide, and their flowers are only 0.1 mm long—or one-tenth the thickness of a dime!
	• However, they seldom flower, but rather reproduce by simple budding.
	• In late summer, some species produce turions, tiny buds that sink, stay on bottom all winter; in spring produce a single gas bubble that carries them to surface.
Warnings, uses	• A great meal for wildlife, since not only do they get a mouthful of plant food, but also all the tiny critters living and laying eggs on the Duckweed mat.
	• Cultivated in Asia for high protein conten t (20%), more than peanuts or alfalfa.
"Stories"	• Depending on size of species, there may be 100,000 to 2,000,000 Duckweeds per square yard.
	• May be aggressive in ponds due to increased sewage runoff, which fertilizes them.
Do this	• With a finger, gently lift a few plants from the water and admire their delicate structure.
	• Different genera have somewhat different structures (purple under or not, up to ten rootlets or one, floats on or under surface or even under other Duckweed, round or oval, etc.): see if you can find several of these.
	• If you lift a fairly large number, use a magnifier to see if you can find evidence of the presence of other organisms.

For your sketches, notes, observations

| POISONOUS | **Elderberry** |
	Sambucus canadensis
About the names	• <u>Elderberry</u>: "elder" is an Old English name for this plant with its abundant berries. • <u>Sambucus</u>: Latin, from Greek name for an ancient flute (the sackbut), since twigs lent themselves to making the same; <u>canadensis</u> means "from Canada."
Memory aid	• Umbrella-like spread of white flowers like an elder's hair; similar spread of fruits like a parasol to keep sun off delicate elders.
When in bloom	• June-July; fruit July-September.
Interesting growth habits	• Wart-like growths on bark are lenticels, openings into the plant for gas exchange.
Warnings, uses	• CAUTION Unripe berries, leaves, roots, and bark all contain cyanide! • Native Americans used tea to treat everything from ulcers to headaches to constipation. • Berries (rich in Vitamin C) used for generations in jelly, pies, wine; flowers may be dipped in batter and fried for fritters; also recipes for chutney, ketchup, syrup, pickles, vinegar —and on and on. • Large white pith is easily hollowed out; hollow twigs used to make spiles (device to direct Maple sap from tapped hole to bucket); also to make whistles and peashooters (but see below!).
"Stories"	• Evidence of domestic cultivation at European Stone Age sites! • Children have become ill from placing toys made of twigs in their mouths. • If you scrape the bark off upwards, use it for an emetic (causing vomiting); if scraped off downwards, use for laxative (don't forget which is which!) • "An extract composed of the berries greatly assists longevity; indeed this is a catholicum against all infirmities whatever." (1664)
Do this	• Look closely at flowers with a magnifier to see their tiny details; sniff for a nice scent. • Look for the elder borer, a long-horned orange and metallic blue beetle, feeding on flowers. • Crush leaves to test the unpleasant scent (a good memory aid), from the cyanide within.

For your sketches, notes, observations

	Enchanter's Nightshade *Circaea quadrisulcata*
About the names	• <u>Enchanter's Nightshade</u>: derived from the Latin name given by Pliny (23-79 A.D., he who died in the eruption of Vesuvius—among his works was a 37-volume natural history). • <u>Circaea</u>: "of Circe," the enchantress who turned Ulysses' men into pigs; <u>quadrisulcata</u> means "four-grooved," referring to the seed pod.
Memory aid	• The plant has simultaneously a gorgeous name and a quite undistinguished overall appearance. This, together with the very tiny size, yet intricate structure, of both flowers and seeds, might suggest the name: something magic to be found here.
When in bloom	• June-August.
Interesting growth habits	• This weedy-looking plant with its tiny nondescript flowers and seed pods rewards a closer look. • One of the very few two-petaled flowers (though since they are deeply lobed, they look like four).
Warnings, uses	• No warnings, no uses—except appreciation of the cunning structures of this evocatively-named plant, known and described so many, many generations ago.
"Stories"	• "…burdened with the singularly inappropriate name of enchanter's nightshade. There is nothing in their appearance to suggest an enchanter or any of the nightshades." (1900, Mrs. William Starr Dana, writing on wild flowers, rather sourly perhaps)
Do this	• Use a magnifier to examine the tiny flowers with their two miniscule deeply lobed petals, and the tiny oval seed pods with their little hooked bristles and the grooves in their surfaces.

For your sketches, notes, observations

	Evening Primrose
	Oenothera biennis
About the names	• <u>Evening Primrose</u>: opens in the evening. • <u>Oenothera</u>: an ancient Greek name apparently having to do with its resemblance to another plant, eaten as an appetizer before wine (<u>oinos</u>); <u>biennis</u> means "lasting for two years, biennial."
Memory aid	• See the X in the flower center: an X-rated flower that only opens at night!
When in bloom	• June-September.
Interesting growth habits	• First-year growth is a low star-burst of long narrow leaves, each with red spots; flower stalk and flowers appear the second year. • Flower stems are fuzzy and reddish. • Stigma in center of flower is shaped like an X. • Usually open only at sunset or in low light; closes by noon.
Warnings, uses	• Many medicinal uses: for some forms of eczema, anti-clotting agent, migraine, cough medicine, PMS, etc., etc.; none of these uses approved by US FDA, although active compounds are present in the plant. • Native American athletes rubbed the root on muscles for added strength. • Very young roots can be boiled in a couple of water changes for a vegetable like carrots.
"Stories"	• This is one of the native flowers that spread from prairies eastward as land was cleared. • Their seeds can remain dormant in former meadow soil for a hundred years, waiting for a disturbance to open up a new meadow. • May be pollinated (at night) by a large sphinx moth.
Do this	• Sniff open flowers to smell the lemony scent. • Look inside to see the X of the stigma. • In winter, gently remove one of the pretty four-parted flower-like woody seed pods and look within to see the brown seeds in rows. • Look for tiny red spots on the rosette leaves. • One author says, "Flowers bloom after sunset, unfolding before the eyes of those who watch them open." Put this to the test!

For your sketches, notes, observations

	Sweet Everlasting *Gnaphalium obtusifolium*
About the names	• <u>Sweet Everlasting</u> since the flowers appear to persist through the winter, and their scent, when bruised, is powerfully sweet and long-lasting. • <u>Gnaphalium</u>: "soft down" for its felt-like leaves, or possibly "tuft of wool," referring to the flower heads; <u>obtusifolium</u> means "blunt-leaved." • Also called Catfoot, Cudweed, Rabbit-Tobacco, Life-Everlasting, Owl's-Crown.
Memory aid	• See above; the scent is powerfully sweet and seems ever-lasting.
When in bloom	• August-November.
Interesting growth habits	• Yellowish bracts, found below tube flowers, do not expand until plant is in seed; thus during the summer the flowers look as if they have not yet opened. • Flower heads remain aromatic even when dried.
Warnings, uses	• Fresh juice used as an aphrodisiac. • Tea used for sore throat, pneumonia, asthma, flu, rheumatism, tumors, diuretic...and on and on.
"Stories"	• Scent described in different sources as that of tobacco, curry, or maple syrup. • If there is a troublesome ghost about, the smoke of the Sweet Everlasting will annoy it sufficiently to drive it away.
Do this	• The scent from one rub of the flowers on your fingers will last for hours. Decide what it smells like to you! If you are with a group, take a poll; the same scent may be interpreted differently by different people.

For your sketches, notes, observations

	False Solomon's Seal
	Smilacina racemosa
About the names	• Styled "<u>False</u>" because it only looks likes another plant called Solomon's Seal (q.v.); large round leaf scars on the root of the "real" plant are said to resemble the seal of King Solomon; furthermore, the "real" plant was said to be so effective medicinally that the "seals" were supposed to have put on the roots by wise Solomon as a seal of approval. False Solomon's Seal doesn't have these medicinal properties. • <u>Smilacina</u> is a diminutive of Smilax, an ancient Greek name for a genus of similar plants, first used by Pliny (23-79 A.D.); <u>racemosa</u> refers to the arrangement of flowers on the stem in a raceme, an elongated cluster of stalked flowers on a central stem. • Also called Jacob's Ladder, Snake-Corn.
Memory aid	• Think of the large and gracefully arching stem of leaves with its tip lit by a burst of white flowers as a kind of scepter held by King Solomon— though of course the "light" is false and not found on the true Solomon's Seal. • Also, flowers have a distinctive smell.
When in bloom	• April-July; fruit July-August.
Interesting growth habits	• Note parallel veins in leaves, marking this as a member of the Lily family. • A colony of plants may all be pointing in the same direction, trying to follow the sun.
Warning, uses	• Smoke was inhaled for two purposes: to treat insanity, and to quiet a crying child. Perhaps the one led to the other? • Tea from the leaves was used as a contraceptive (perhaps avoiding the above problems in the first place).
"Stories"	• It's hard for seeds to reach the ground or germinate through the thick layer of slowly-decaying mulch in oak forests, so once they do, plants often spread by underground runners. The size of a colony of False Solomon's Seal might therefore be related to the age of the woods.
Do this	• Smell flowers to fix identifying scent in mind.

For your sketches, notes, observations

POISONOUS	**Bracken Fern** *Pteridium aquilinum*
About the names	• <u>Bracken</u> comes from an Old Norse word for "fern." • <u>Pteridium</u> means "small fern" (although it isn't); the <u>pteri</u> part comes from the Greek for "feather," referring to ferns; <u>aquilinum</u> means "eagle-like," referring to the appearance of the unfolding fiddleheads, which look like eagle talons. • Also called Brake-fern.
Memory aid	• Coarse and common, with three big fronds: visualize three coarse green toads standing in a circle and croaking "Braaack, braaack, braaack" at each other
When in fruit	• Fiddleheads in early spring (among the earliest); fruit dots in late summer.
Interesting growth habits	• Found on every continent except Antarctica. • Indicates poor soil. • Produces new, irregularly-formed leaves all season; killed by first frost. • Rootstock may extend for ten feet underground. • "Our most common fern."
Warnings, uses	• CAUTION Carcinogenic; contains compounds used to form tumors in experiments. • NOTE OF CAUTION Never eat any fiddleheads the species of which you are not positive. • Poisons livestock; contains substance which interferes with processing of B vitamins. • Native Americans used a wash of it to promote hair growth; tea used to treat stomach cramps.
"Stories"	• If you burn a Bracken Fern, it will rain; an early English king, on vacation, forbid the burning of Bracken on his visit, and the weather, lo and behold, was sunny! • Also, if you need to protect against witches and goblins, this is your plant. • Finally, to discover the first letter of your future spouse's name, just cut the stem of a Bracken Fern and there it will be: C.
Do this	• Fertile fronds have rolled edges; find one, turn it over and uncurl the edges to find the rows of fruit dots.

For your sketches, notes, observations

	Christmas Fern *Polystichum acrostichoides*
About the names	• Green at Christmastime. • Polystichum means "many rows" (for the arrangement of its fruit dots); <u>acrostichoides</u> means "rows across the top" (the fruit dots are found only on the highest leaflets).
Memory aid	• Each leaflet looks like a Christmas stocking, with leg and foot.
When in fruit	• Evergreen; fruit dots ripe June -October.
Interesting growth habits	• As on many other evergreens, leaves are leathery. • Often found on rocky slopes; likes limy soil. • Fronds with fruit dots (fertile) are taller, more rigid, and more erect than those without fruit dots (sterile). • Only sterile leaves are evergreen.
Warnings, uses	• Helps control erosion on bare slopes. • Used by early New Englanders for Christmas decoration.
"Stories"	• When the baby Jesus was born, all the flowers that made up the straw that cradled Him in the manger burst into flower; only the fern forgot to bloom, and so it has not been able to flower since.
Do this	• Check a cluster of Christmas Ferns to see if the taller fronds really are the fertile ones, with fruit dots in rows on the back of the highest leaflets. • Feel the leaves to see how leathery they are; this quality helps them to retain their moisture during the winter.

For your sketches, notes, observations

| | **Cinnamon Fern** |
	Osmunda cinnamomea
About the names	• <u>Cinnamon</u>: color of woolly tufts at bases of leaflets, and of spore cases, withered fertile stalks, and older fiddleheads.
	• <u>Osmunda</u>: either see story below; or possibly from the Anglo-Saxon god of thunder, Osmunda; or from <u>os</u> ("bone") + <u>mundare</u> ("to clean") for its medicinal uses named by Linnaeus in 1753; <u>cinnamomea</u> means "cinnamon-brown."
Memory aid	• Just check underneath frond for the tufts of cinnamon-colored hairs in each axis, or look for the "cinnamon-stick"-like fertile stalks.
When in fruit	• Early spring.
Interesting growth habits	• Cinnamon-colored fertile stalks appear first, before more obvious lacy fronds; fertile stalks bright green at first, than turn cinnamon color.
	• May grow over five feet tall; "sometimes so profuse as to form jungle-like areas."
	• Spores must germinate quickly, or die within a few days.
	• Young fiddleheads are silvery-hairy.
	• Spores are green instead of brown like other fern spores.
Warnings, uses	• NOTE OF CAUTION Not all spring fiddleheads are edible, so don't experiment! However, those of Cinnamon Fern can be cooked and eaten.
	• The central ball of roots and rhizomes, called the Heart of Osmund or Bog Onion, can be cooked and eaten early in spring.
	• This same mat of rhizomes is used in greenhouses as a medium for germination of orchids.
"Stories"	• The rootstocks renew themselves at one end each year while dying off at the other; "It is believed they can live forever," botanist Broughton Cobb is quoted as saying.
	• "A certain Osmund, living at Loch Tyne, saved his wife and child from the inimical Danes by hiding them...among masses of flowering ferns;...the child so shielded named the...plants after her father."
Do this	• Look in leaflet axils for tufts of cinnamon-colored hairs.
	• Look around for separate fertile stalks.

For your sketches, notes, observations

	Hayscented Fern
	Dennstaedtia punctilobula
About the names	• <u>Hayscented</u>: fronds smell like hay when crushed. • <u>Dennstaedtia</u>: Augustus Dennstaedt was a German botanist of the early 19th c.; <u>punctilobula</u> means "dotted-lobed."
Memory aid	• The hay scent is unmistakable—very sweet and fresh—when the fronds are crushed.
When in fruit	• May-September; fruit dots ripe in August-September.
Interesting growth habits	• May grow in colonies that exclude everything else. • Fronds have tiny glands on them (containing the scenting substance); when crushed this substance can be felt on the fingers as lightly tacky. • Fronds are hairy on the undersides. • Fronds tend to become ragged by late summer. • Fronds tend to follow the light during the day.
Warnings, uses	• NOTE OF CAUTION Not all fiddleheads (earliest fern sprouts) are edible; some are poisonous. Do not eat fiddleheads unless gathered by an expert!
"Stories"	• Before the discovery of spores, people believed that ferns had seeds—very very tiny ones. On Midsummer's Eve, if you went out at midnight, you were supposed to be able to catch a fern seed when it became ripe. This seed was magic; in Shakespeare's <u>Henry IV</u> it rendered one invisible.
Do this	• Crush, sniff, and feel the fronds. • Look with a magnifier at the tiny hairs.

For your sketches, notes, observations

	Interrupted Fern *Osmunda claytoniana*
About the names	• <u>Interrupted</u> refers to interruption of ranks of sterile leaflets by fertile leaflet s. • <u>Osmunda</u>: for Anglo-Saxon god of thunder, Osmunda; or from <u>os</u> ("bone") + <u>mundare</u> ("to clean") for possible medicinal uses; or for Osmund the waterman, who hid his wife and child amid the ferns during an attack. Named by Linnaeus in 1753, and you can tak e your pick of these reasons or make up your own; <u>claytoniana</u> for John Clayton, who discovered this fern in Virginia in the early 1700's.
Memory aid	• Interruptions along the frond of leaflets, the interruptions either empty or lined with brown fertile leaflets.
When in fruit	• One of earliest ferns to appear in spring; fertile leaflets gone by early summer.
Interesting growth habits	• One of largest ferns in forest, in clusters four to five feet tall. • Fiddleheads appear very early and are very woolly and brown.
Warnings, uses	• NOTE OF CAUTION Some fiddleheads are not good to eat, so never experiment. • Especially do not experiment with the fiddleheads of Interrupted Fern, which, with their woolly coats, have "all the taste appeal of green hairballs."
"Stories"	• "An especially good subject for the woodland garden."
Do this	• Look around to try to find a frond which is interrupted by fertile leaflets, in early spring; or with empty space in center of stalk where fertile leaflets have been and have withered, by summer. Note: fertile stalks are taller than sterile ones. • Remember: if it has brown woolly tufts under, it's a Cinnamon Fern, not an Interrupted Fern!

For your sketches, notes, observations

	Lady Fern
	Athyrium Filix-femina
About the names	• Reason for English name uncertain. • <u>Athyrium</u> means "lacking a door," referring to the fact that the covering of the tiny spore cases opens late (note that another source says this word means "sporty," referring to the varied shapes of the spore cases: in biology a sport is a dramatic change, as if from a mutation). • <u>Felix-femina</u> just means "female fern."
Memory aid	• Observe: very lacy and delicate, with tips of fronds drooping modestly, the perfect lady (not to be confused with the Hay-Scented Fern, which looks a lot like it, but of course a Lady would never have that grassy odor, or that sticky feeling of the Hay-Scented!).
When in fruit	• Early summer.
Interesting growth habits	• Described as "discouragingly variable." • Spore-cases can be rectangular, chevron-shaped, horseshoe-shaped, curved—or what have you.
Warnings, uses	• NOTE OF CAUTION Some fiddleheads are not good to eat, so do not experiment. • Native Americans used a tea made of the stem to ease labor pains, and a tea of the root to induce milk, and a powder of the roots for sores.
"Stories"	• Wonderful to contemplate that the ancestors of ferns appeared on Earth 375 million years ago!
Do this	• Look for the lax tips, then check for scent and sticky feel to tell a Lady Fern from a Hay-scented Fern.

For your sketches, notes, observations

	New York Fern
	Thelypteris noveboracensis
About the names	• <u>New York</u>: "Linnaeus received this fern from Canada, and never explained why he used for its epithet a latinization of New York." • <u>Thelypteris</u>: "female-fern" (why? possibly because they are small and delicate for the most part? The name dates to 1762); <u>noveboracensis</u> meaning "New York."
Memory aid	• Leaves are tapered at each end: New Yorkers burn the candle at both ends.
When in fruit	• Fruit dots appear in summer.
Interesting growth habits	• Fertile leaves are taller, narrower, and more upright than sterile. • Especially found in open, sunny areas in woods, where there may be quite large colonies of them. • Pale yellow-green. • Leaves may turn to follow the sun during the day.
Warnings, uses	• NOTE OF CAUTION Not all fern fiddleheads are edible; be sure of identification before eating.
"Stories"	• All ferns reproduce in the ancient manner of the first plants: spores, if they land on moist ground , develop into a miniscule plantlet with both sexes, which under exactly favorable circumstances will fertilize itself and then grow into the fern we see.
Do this	• Look for fertile leaves (with a few fruit dots on back) to see if indeed they are taller, na rrower, and more upright than sterile ones. • Note whether all leaves are facing the sun.

For your sketches, notes, observations

| | **Polypody Fern or Rock-Cap Fern** |
	Polypodium vulgare
About the names	• <u>Rock-Cap</u> from its habit of growing on top of rocks. • <u>Polypodium</u> means "many feet," referring to the runners by which the fern spreads in a mat; <u>vulgare</u> means "common."
Memory aid	• Easy to remember since it so often is found capping rocks.
When in fruit	• Evergreen; fruit dots ripen in summer.
Interesting growth habits	• May be found in about ten different forms, each named for the varying appearance of the leaflets. • A nice sight in winter amid the snow, as a colony rises upright from boulders, unlike some other evergreen ferns, which are green but prostrate in winter. • Fruit dots are big and prominent on upper leaflets; they may be seen in winter too.
Warnings, uses	• Helps to form and stabilize soil on slopes.
"Stories"	• Sometimes sold as Resurrection Fern: without water plants curl up and look dead, but with addition of water they rise and become green again. • Thoreau rejoiced in the "fresh and cheerful communities" of the Rock-Cap Fern in the springtime.
Do this	• Turn fronds over to see the big fruit dots, which in young plants may be golden-yellow.

For your sketches, notes, observations

	Royal Fern *Osmunda regalis*
About the names	• <u>Royal</u>: well, this is a large and impressive fern, rather elegant in form, but no more than some others; no real explanation is found for its common name. • <u>Osmunda</u>: from the Anglo-Saxon god of thunder, Osmunda; or from <u>os</u> ("bone") + <u>mundare</u> ("to clean"), for its supposed medicinal properties; or for a certain Osmund, a waterman, whose wife and children hid among these ferns during an invasion; <u>regalis</u> means "outstanding, regal." • Also called Flowering Fern, on account of the appearance of its fertile stalks.
Memory aid	• The royal king carries a scepter (the greenish later brownish, fertile stalk rising from some plants).
When in fruit	• Fertile stalks develop in April, ripen in mid-summer.
Interesting growth habits	• The fertile stalk, with its spore cases clustered like a flame at the tip, rises from the center of the plant, the infertile leaves at either side of the same stalk. • Unusual amount of empty space between leaflets on a leaf and between the leaves on the stalks. • May grow to five or six feet. • Nearly always found very near wetlands.
Warnings, uses	• Supposed to be good for treating jaundice, rickets, injuries caused by falling from high places, lumbago, colic, and ruptures and burstings (after the fall from a high place?).
"Stories"	• Also sometimes called Locust Fern, because the leaves rather resemble the compound leaves of that tree. • One guide says, disparagingly, "By some it is considered our finest fern; actually it is too coarse really to reflect typical aesthetic characteristics of ferns, with their lacy, light-filtering and delicate growth and form." Strictly eye of the beholder! • If it's growing in plenty of water, it may get to resemble tropical growth.
Do this	• Use a magnifier to examine the fertile part of the plant.

For your sketches, notes, observations

	Sensitive Fern *Onoclea sensibilis*
About the names	• <u>Sensitive</u> because it turns brown and withers with the first frost. • <u>Onoclea</u> means "closed cup," referring to fruit dots covered by curled-over edges of their covers; <u>sensibilis</u> means "sensitive to touch," although these aren't. • Also called Bead Fern.
Memory aid	• Edges of leaflets are wavy, like S's for Sensitive. Has <u>s</u>eparate <u>s</u>pore <u>s</u>talks (S for Sensitive). Even though the leaves are large and sturdy-looking, they keel over with first frost, they're so sensitive.
When in fruit	• Spores form in late summer/fall, released in spring.
Interesting growth habits	• Spores are carried on separate stalks; their capsules look like rows of brown beads. If you open one, it is full of brown "dust," which is actually thousands and thousands of spores. • These fertile stalks last all winter and on into the next and possibly the one after. • Fiddleheads are pale red, in a mass in spring. • "Sturdy, coarse, unfernlike fern."
Warnings, uses	• NOTE OF CAUTION Not all fiddleheads are edible. Don't eat any unless you are certain of them.
"Stories"	• This plant has a front and a back: spore cases are arranged on only one side of the fertile stalk.
Do this	• Gently shake the fertile stalks to empty some dust-like spores onto your hand. • Look for fields of the free-standing brown fertile beaded stalks in the fall and winter. • Look closely to see if spore cases have opened or are still closed.

For your sketches, notes, observations

CAUTION	**Daisy Fleabane**
	Erigeron annuus

About the names	• <u>Daisy</u>: looks like a daisy (but isn't); <u>Fleabane</u>: if you put the dried flowers around your house, the fleas will disappear (it's said). • <u>Erigeron</u> means "early old man," presumably because it has a kind of white beard and blooms in spring, so named by Theophrastus (c. 370 - 285 B.C.; he studied plants brought back from the campaigns of Alexander the Conqueror); <u>annuus</u> means "annual" (as opposed to perennial). • Also called Lace-buttons, Little-Daisies, Sweet Scabious.
Memory aid	• It looks like a daisy, but the "petals" are as tiny, and numerous, as fleas.
When in bloom	• May-October (one source says it begins late March!).
Interesting growth habits	• Fifty to one hundred tiny ray flowers crowd around the cluster of yellow disk flowers. • These look like Asters, but Asters are perennial and fall-blooming. Daisy Fleabane is an annual and begins blooming in spring.
Warnings, uses	• CAUTION Skin may be irritated by contact. • Used to stop various kinds of hemorrhages; also to treat urinary tract problems.
"Stories"	• This plant is an alien here in New England, for it migrated here from the prairies to the west, as our New England forests were gradually cleared during Colonial days, providing new habitats for open-field plants.
Do this	• Try counting all those little bitty rays!

For your sketches, notes, observations

Gall-of-the-Earth or Rattlesnake Root
Prenanthes trifoliata

About the names	• <u>Gall-of-the-Earth</u>: a mysterious name dating from at least 1567, referring to a plant's bitterness, though not this plant, but rather one said to have been discovered by Chiron the centaur, a physician of Greek mythology. • <u>Rattlesnake Root</u> refers to the plant's supposed efficacy against rattlers. • <u>Prenanthes</u> means "drooping flower," a perfect description; <u>trifoliata</u> describes the three-parted leaves.
Memory aid	• The whole aspect of the plant—its fall bloom time, waxy, drooping flower clusters that seem to hug close to the stem, its unexpected appearance by the side of the trail, not showy but obvious when you finally see it—somehow suggests the melancholy name, evoking the sadness, bitterness, gall of life on earth.
When in bloom	• September-October.
Interesting growth habits	• The leaf shape is highly variable. • Though these are composite flowers, with both ray and disk flowers, unlike most other composites, they look down, away from the sun. • Stems contain milky sap.
Warnings, uses	• Native Americans used root to stimulate milk flow after childbirth. • Good for snakebites and dogbites (see story below for why this dual use might come in handy).
"Stories"	• "The rattlesnake has an utter antipathy to this plant, insomuch that if you smear your hands with the juice of it, you may handle the viper safely. [Do Not Try This At Home] Thus much can I say of my own experience, that once in July, when these snakes are in their greatest vigor, I besmeared a dog's nose with the powder of this root, and made him trample on a large snake several times, which, however, was so far from biting him, that it perfectly sicken'd at the dog's approach, and turn'd its head away from him with the utmost aversion." (William Byrd, 1728) Pretty tough on the dog, not to mention the snake.
Do this	• Tip flower up; examine parts with a magnifier.

For your sketches, notes, observations

| | **Garlic Mustard** |
	Alliaria officinalis
About the names	• <u>Garlic Mustard</u>: it smells like garlic and is in the Mustard family. • <u>Alliaria</u>: "garlic-smelling"; <u>officinalis</u>: "of the apothecaries" (sold as medication). • Also called Jack-by-the-Hedge, Sauce-Alone.
Memory aid	• The leaf smell is distinctly that of <u>garlic</u>; combine that with its seed pods which run up the stem to escape the hot <u>mustard</u>.
When in bloom	• April-June.
Interesting growth habits	• Highly invasive and aggressive, leafs out early and outcompetes native spring-blooming flowers. • Overwinters as green rosette. • Long, thin seed pods lying close to the stem and pointing upward, typical of the Mustard family.
Warnings, uses	• If your cow eats it, her milk will taste of garlic. • Used in old times for gangrene and ulcers, and in a syrup with honey for edema. • Also in old times, country folk made a sauce of the plant to be eaten with bread and butter, or raw in a salad (which was said to "warm the stomach" and aid digestion).
"Stories"	• "The seeds, when snuffed up the nose, excite sneezing." • One plant can produce up to 6,000 seeds; colonies have been shown to contain 20,000 seedlings in one square meter.
Do this	• Rub leaves, sniff, and think of garlic bread. • Open a pod to look for the 3 mm seeds.

For your sketches, notes, observations

| | **Wild Geranium** |
	Geranium maculatum
About the names	• <u>Geranium</u>: "crane" (given by Dioscorides, c. 64 A.D.), referring to shape of fruit; <u>maculatum</u>: "spotted," possibly for blotchy coloring of flowers. • Also called Cranesbill.
Memory aid	• Little resemblance to hothouse "Geraniums"! Five-parted leaves (and stem) are hairy -<u>Ger</u>-ry; 5 letters in hairy and gerry and 5 parts to leaf.
When in bloom	• April-June; fruits June-July.
Interesting growth habits	• Long beak, in center of flower, looks like a crane's bill as it enlarges into seed pod. • Single flower lasts only one-three days. • Flower may be in male or female phase, depending on weather.
Warnings, uses	• Underground stem has high tannin content (10 - 20%) and so is highly astringent; used to stop bleeding, especially on sores inside mouth. • Also used for dysentery and diarrhea. • Attracts Japanese beetles, and its fragrance has been used as bait in Japanese beetle traps.
"Stories"	• When long seed pod dries, it pops open and shoots seeds away by means of five curled strips which join at base of seed pod.
Do this	• Examine flowers in spring and summer for long crane's bill center; in summer and fall for five curling parts of open pod. • Look for darker purple or translucent "nectar guide" lines, leading to center of flower. • Look for color variations in a stand of these flowers; there are rare pure white forms as well as variations into rose and purple. • Look to see if male anthers are still on flower or if flower has converted to female stage and anthers have dropped off.

For your sketches, notes, observations

	Goldenrods
	Solidago species
About the names	• <u>Goldenrod</u> describes the rod-shaped clusters of bright yellow flowers (note that there is a white member of the genus, called, of course, Silverrod). • <u>Solidago</u>: a uniter (to make solid), since it heals up wounds.
Memory aid	• The name is perfectly descriptive.
When in bloom	• July-October (depending on species).
Interesting growth habits	• About 100 species in U. S., 75 in our area! • "Distinguishing the various species...often perplexes even the trained botanist"; as a start, there are five shapes the flower cluster may take. • Many insects lay their eggs within different Goldenrods, which then produce galls; knowing which insect prefers which species of Goldenrod can help in identification. • Each flower head may contain 1,000 flowers.
Warnings, uses	• Used before and during the Crusades to heal wounds, and Crusaders brought it back to Europe. • Used for ulcers, loose teeth, fever, etc. etc. • Flowers cooked as fritters or for tea. • Flowers make a fine yellow dye.
"Stories"	• Goldenrod is NOT responsible for fall hayfever! Its flowers are brightly-colored to attract insect pollinators, thus it wants to hold onto its pollen; plants with drab flowers (such as Ragweed) are wind-pollinated (no need to attract insects). But Goldenrod and Ragweed bloom at the same time, and Goldenrod is so highly visible that it gets blamed for all that pollen floating in the air and into your nose. • At the time of the Boston Tea Party, the Colonists made "liberty tea" from Goldenrod. • England bought lots of Goldenrod from the Colonies, but once it was discovered in England, the bottom dropped out of the market.
Do this	• Search diligently among the flowers to find the beautiful Goldenrod Spider, blending right in; the Goldenrods host many wonderful insects. • With a magnifier, look closely at an individual flower to see both its disk and ray flowers.

For your sketches, notes, observations

	Goldthread *Coptis groenlandica*
About the names	• <u>Goldthread</u>: a lovely name that is perfectly descriptive of the delicate deep yellow roots. • <u>Coptis</u>: "cut," referring to the divided leaves; <u>groenlandica</u>: "from Greenland" since it is found there. • Also called Canker Root, Mouth Root.
Memory aid	• Although it has "leaves of three," like Poison Ivy, its leaves are festive--dainty, scalloped, and strawberry-like, and it is a low ground-cover quite unlike PI; identify it by looking at the golden thready roots.
When in bloom	• May-July.
Interesting growth habits	• Flowers and leaves on separate stems. • Evergreen. • Likes cool, mossy habitats, where it grows in mats.
Warnings, uses	• Roots steeped in hot water were used as a mouthwash to heal canker sores. • Decoction of Goldthread and Goldenseal supposed to be useful in reducing craving for alcohol. • Roots used as a yellow dye.
"Stories"	• The dried roots went for a dollar a pound in the 1880's; a pound of roots is a lot of roots and a lot of labor!
Do this	• Gently pull up one plant to see the gold-colored threads of its roots. • Look closely at the flower: what look like white petals are actually sepals, and the petals are reduced to small clublike structures.

For your sketches, notes, observations

	Grasses
	Gramineae species
About the names	• <u>Grass</u> comes from an Old English word which is also the root of the words "green" and "grow." • <u>Gramineae</u>: from the Latin for "grass."
Memory aid	• "Sedges have edges; Rushes are round. Grasses have joints (when the cops aren't around)."
When in bloom	• Any time from April-October; fruits (grain) summer-fall/winter.
Interesting growth habits	• Stems are hollow between joints; each joint is closely wrapped by a sheath which is the base of the leaf, one leaf to a joint. • Branching ("tillering") arises at the base of the plant, so it often makes a rosette or "tussock"; this allows plant to branch even when trod on, grazed or burned to the ground. • Found everywhere, even in Antarctica. • It has a special type of photosynthesis which enables it to use only about half as much water as other plants need to produce an equal amount of plant material.
Warnings, uses	• Sugar, wheat, rice, rye, oats, corn, barley, millet, sorghum, bamboo... for building and thatching, for making baskets, oil for cooking, food for birds...and on and on.
"Stories"	• For humans, the most important of the plant families, since it provides not only all of the grains we eat, but also those eaten by our domestic animals. • The world's most widely distributed plant family with the largest number of individuals. • What you see may be only 10% of the total weight of the plant; the rest is in its root system, thus conserving water during drought. • A 4-month old rye plant grown in a greenhouse was found to have 387 miles of roots!
Do this	• In spring and summer, use a magnifier to examine the complex, tiny flower structures. • In fall, look for the heads of grain—all grasses have them! • Also look for the little hairs that form a ring around the juncture of the sheath and the stem. • How many different kinds of grass can you find in a small meadow? Try sketching them.

For your sketches, notes, observations

Greenbrier
Smilax rotundifolia

About the names	• <u>Greenbrier</u>: from an Old English word meaning "brier," derivation unknown, plus the color. • <u>Smilax</u> from an ancient Greek name for Bindweed, a kind of vine; <u>rotundifolia</u> is the leaf shape. • Other very similar species, or this species, may also be called Catbrier. • Other names are Bullgrip, Dogbrier, Horsebrier, Devil's-Wrapping-Yarn, Wait-a-Bit.
Memory aid	• Could not be called otherwise! with its thorny twisted tangles of green stems and leaves.
When in bloom	• April-August; fruits September-spring.
Interesting growth habits	• Only woody vine in Northeast with both thorns and tendrils; stems have angles or facets. • Blue-black berries in clusters; inside, seeds enclosed in a stretchy membrane. • Tendrils attached at each end coil where attached, forming a straight section between two coils of alternate directions.
Warnings, uses	• Best to try to avoid getting tangled in it. • However, it provides a great safe hiding place for small animals and birds, including catbirds, rabbits, deer. • "From these roots while they be new or fresh beeing chopt into small pieces and stampt is strained with water a juice that makes bread, and also being boiled a very good spoonemeat in maner of a gelly, and is much better in taste if it bee tempered with oyle." (1590, Thomas Harriot, English scientist who went to Virginia in 1584 in the attempt to colonize at Roanoke.) • Native Americans rubbed prickles on their skin as a counter-irritant to relieve cramps and twitching; tea from roots was used to help expel afterbirth in delivery. • Young shoots may be cooked like asparagus.
"Stories"	• "Flexible rolls of vegetative barbed wire."
Do this	• At small-animal level peer inside the brierpatch to find runs and trails, and to imagine hiding in there. • Look for fruit; squeeze seeds in fingers and slowly pull apart to see the stretchy membrane. • See intricate twists of tendrils.

For your sketches, notes, observations

| | **Orange and Yellow Hawkweeds** |
	Hieracium aurantiacum/pratense
About the names	• <u>Hawkweed</u>: Pliny of ancient Greece said that hawks swoop down and suck the sap of this plant, which was supposed to improve both their eyesight and that of humans.
	• <u>Hieracium</u> comes from the Greek word for "hawk"; <u>aurantiacum</u> means "orange-colored" from the word for "gold"; <u>pratense</u> means "of meadows."
	• Orange Hawkweed also called Artists'-Brush, Redweed; Yellow Hawkweed also called King Devil.
Memory aid	• Entire plant covered with stiff hairs, like bristles: visualize a fierce hawk swooping down on it, his wings, talons, and tail taut and sticking out in bristly fashion.
When in bloom	• June-September (Orange); May-August (Yellow).
Interesting growth habits	• Note that there are other species of Hawkweeds similar in appearance to these; these two common ones are alien and have leaves in basal rosettes, whereas native species' leaves are found along the stem.
	• Member of the Composite family, which usual ly have both disk and ray flowers; but Hawkweed has only ray flowers (like a dandelion).
Warnings, uses	• Often described as "a troublesome weed."
"Stories"	• Because "the stalkes and cups of the floures are all set thick with a blackish downe or hairinesse as it were the dust of coles...the women, who keep it in gardens for noveltie sake, have named it Grim the Collier [dealer in coal]." (1633) It seems that "Grim the Collier" was the name of a popular English comedy of the time.
Do this	• Are the leaves of your Hawkweed arrayed along the stem (a native species) or in a basal rosette (alien)?
	• Examine the stiff hairs with a magnifier to see the tiny glands on their tips.

For your sketches, notes, observations

| | **Heal-All** |
	Prunella vulgaris
About the names	• "Students of plant lore are at a loss to explain how <u>Heal-All</u> got its name, for its medicinal uses were always limited." • <u>Prunella</u> comes from a German word for "brown," which described symptoms of a virulent illness passing through imperial troops in 1547 and 1566; the illness was treated with Heal-All; <u>vulgaris</u> means "common." • Also called Selfheal, Carpenter's-Weed, Blue-Curls, Heart's-Ease.
Memory aid	• Flower head cylindrical, like a pill bottle, doubtless filled with pills to heal all ills.
When in bloom	• May-September/November.
Interesting growth habits	• Although a Mint, with a square stem, it has no minty smell. • A very nondescript plant, until you look closely at the flowers.
Warnings, uses	• Considered a weed, may invade lawns. • Plant contains a number of important medicinal compounds: antibiotic, anti-tumor, diuretic, hypotensive, and antimutagenic. So maybe Heal-All is a good name, after all.
"Stories"	• Robert Frost's poem "Design": I found a dimpled spider, fat and white, On a white Heal-All, holding up a moth Like a white piece of rigid satin cloth— Assorted characters of death and blight Mixed ready to begin the morning right, Like the ingredients of a witch's broth— A snow-drop spider, a flower like froth— And dead wings carried like a paper kite. What had that flower to do with being white, The wayside blue and innocent Heal-All? What brought the kindred spider to that height, Then steered the white moth thither in the night? What but design of darkness to appall?— If design govern in a thing so small.
Do this	• By all mean examine the lovely tiny flowers with a magnifier. • Feel the square stem of the Mint family.

For your sketches, notes, observations

	Hemlock
	Tsuga canadensis
About the names	• <u>Hemlock</u>: from the old word for Poison Hemlock (a quite different genus), derivation unknown. • <u>Tsuga</u>: from the Japanese name for the Hemlock Cedar (Hemlock trees are native both to North America and Asia); <u>canadensis</u>: "from Canada."
Memory aid	• Needles have tiny stalks: Hemloc <u>K</u> stal<u>K</u>.
When in bloom	• May-June.
Interesting growth habits	• Pink or pale green female flowers like tiny cones at twig ends; yellow male flowers very tiny, found along twigs. • Those tiny mature cones, which appear when tree is about 20 years old, are as big as they get. • Chemicals in shed needles may prevent other plants from growing under the tree. • Grows slowly in low light of parent; 40-60 year old seedling may be only six feet tall.
Warnings, uses	• Tea from young needles is rich in Vitamin C. • Inner bark can be used "in time s of emergency" to make a kind of flour; "the taste…is unattractive but nourishing" (probably better to check your food supplies before you go out). • Bark is astringent and used to stop bleeding. • Used in leather tanning process, especially in 19th century when big sheets of bark were pulled off trees, leaving them to die.
"Stories"	• Hemlock bark is so thick that it may be twenty percent of the total volume of the tree. • Snow accumulation under a big tree is much reduced; deer may gather there in winter. • "Has produced more interesting variant forms than any other native tree…horticulturalists have named perhaps one hundred of these." • Oldest recorded Hemlock was 988 years old. • Can survive cold of 100 degrees below zero.
Do this	• Are there are twig tips and short branches on the ground under a Hemlock? This may be a sign of porcupines feeding high in the tree. • Look on branches for a white, cobwebby substance; this is sign of the Hemlock Wooly Adelgid, a new pest which is killing off our Hemlocks (check your home trees too).

For your sketches, notes, observations

	Shagbark Hickory
	Carya ovata
About the names	• <u>Shagbark</u> is perfectly descriptive! <u>Hickory</u> from an Algonquin word "pocohiquara," a drink made from pressed hickory nuts. • <u>Carya</u>: "walnut" in ancient Greek (Hickory is in the Walnut family), derived from a word describing nuts with hard shells; <u>ovata</u> means "egg-shaped," referring to leaves, specifically leaves whose broader end is toward the twig.
Memory aid	• Hickories have compound leaves, leaflets in sets of 5 or 7; the odd, and larger, leaf is at the end of each stem, like a lollipop on a stick: think lickory-hickory; Shagbarks have shaggy bark.
When in bloom	• Catkins in April; nuts in September/October.
Interesting growth habits	• Younger trees have smooth bark, by age forty the bark begins to curl in long strips away from the trunk, making one long for a big comb to tidy the tree! • Nuts (borne after about forty years) enclosed in extremely thick, hard husk; splits in four when ripe; big crops every three years. • May live two to three hundred years.
Warnings, uses	• Nuts eaten raw, ground into flour, or as oil source; Algonquins pounded them, then steeped them in boiling water, for a kind of nut-milk. • Pioneers used inner bark for yellow dye. • Wood used to produce hickory-smoking of food. • Firewood and charcoal: a cord of hickory wood said to give as much heat as a ton of coal or 175 gallons of fuel oil! • Wood is very hard yet elastic; used for tools, skis, baseball bats, bows and arrows, gunstocks, furniture.
"Stories"	• General, then President, Andrew Jackson was called "Old Hickory" by his troops on account of his toughness.
Do this	• Use a magnifier to see tufts of hair on teeth of leaflets.

For your sketches, notes, observations

	Hog Peanut *Amphicarpa bracteata*
About the names	• <u>Hog Peanut</u>: A member of the peanut family; its underground fruits may be rooted up by hogs. • <u>Amphicarpa</u> means "both kinds of fruit," because this plant forms fruit from its two kinds of flowers; <u>bracteata</u> means "bracted" (referring to bracts, tiny leaves at the base of the flower). • Also called Goober (derived from an African word for peanut, nguba), Peavine, Wild-Peanut.
Memory aid	• Leaves in threes, like three letters in <u>hog</u> and <u>pea</u>. • Tends to <u>hog</u> other plants by twining and climbing over them.
When in bloom	• August-September.
Interesting growth habits	• Showy light purple pea-type flowers hang in clusters above; tiny flowers without petals grow on lower or creeping branches. The showy flowers are cross-pollinated by insects, but if for some reason this does not happen, the tiny inconspicuous flowers, which never open, pollinate themselves. This is called cleistogamy, which means "hidden marriage." Violets, among others, have the same system.
Warnings, uses	• Fruits of lower flowers are underground pods with an edible seed in each, to be cooked like a bean, seasoned with butter, salt and pepper.
"Stories"	• "...a very important food plant among the American Indians...the women in autumn and early winter robbed the nests of...rodents, securing big piles of [the nuts}. The Dakota Nations when taking seeds from the nests of animals left corn or other food in exchange." (1939)
Do this	• Search for the two kinds of flowers on one plant.

For your sketches, notes, observations

| | **Huckleberries** |
	Gaylussacia species
About the names	• <u>Huckleberry</u>: probably a corruption of "hurtleberry" (for the same plant)—but its derivation is unknown. • <u>Gaylussacia</u>, named for French chemist J. L. Gay-Lussac, 1778-1850 (discovered boron and the laws of gas under pressure); he was a professor at the Sorbonne and at the Jardin des Plantes in Paris. A yellowish mineral is named for him—perhaps species name refers to the yellowish resin dots on its leaves?
Memory aid	• Spar<u>Kl</u>y sti<u>CK</u>y hu<u>CK</u>leberry, with its resin dots.
When in bloom	• May-June; fruit July-September.
Interesting growth habits	• Huckleberry plants look a lot like Blueberries, but they have shiny resin dots on the leaves (some species on both sides of the leaf, others only underneath). • Bark is dark and smooth down to the ground; new twigs are red (cf. Blueberry, with its rough bark and new twigs which are green and red, or just green).
Warnings, uses	• Some people like its fruit better than Blueberries; can be used raw or cooked or in jelly. • Ground-walking birds—grouse, bobwhite, turkey, mourning dove—eat the fruits.
"Stories"	• The derivation or meaning of "hurt" in hurtleberry/huckleberry is unknown; it doesn't mean "to injure"—make up your own exotic story!
Do this	• Use a magnifier to see the shiny little resin dots on the leaves. • One naturalist (Stokes) suggests pressing the underside of a leaf on the back of one's hand, to see the yellow mark left by the resin dots. • If you can find a fruit, open it up to count the ten seedlike nutlets.

For your sketches, notes, observations

| POISONOUS | **Indian Pipe** |
	Monotropa uniflora
About the names	• <u>Indian Pipe</u>: the shape of the flower plus stalk. • <u>Monotropa</u>: "one turn of the stem," possibly referring to the pronounced droop at the tip of the flower stalk; <u>uniflora</u> means "one flower," and that is all each plant makes. • Also called Corpse Plant, Ghost Flower, Ice Plant.
Memory aid	• Stalk plus flower = shape of a pipe.
When in bloom	• June-September.
Interesting growth habits	• Has no chlorophyll, so cannot photosynthesize; instead, as a saprophyte, it must hook its roots up with underground fungi to get nourishment transferred from decaying plant material or from nearby green plants to which the fungus is connected. • Can grow in deep shade since it doesn't photosynthesize. • Flower nods until pollinated, when it turns up, forms a seed capsule, and turns black. • In winter it persists as a forlorn cluster of black wires with a dab of pod at their tips.
Warnings, uses	• CAUTION Contains several toxic compounds. • Native Americans may have used its juice as an eye lotion—and to treat bunions. • Contains anti-bacterial compounds.
"Stories"	• One writer says, "This subfamily...is in need of considerable study...many details of their structure and relationships are in doubt." (Heywood, 1993) • Not known how it is pollinated. • "...deserves its name of corpse plant, so like is it to the general bluish waxy appearance of the dead; then, too, it is cool and clammy to the touch, and rapidly decomposes and turns black even when carefully handled." [shudder] (1887, Dr. Millspaugh, physician and curator of botany at the Field Museum in Chicago).
Do this	• Are flowers turned up (pollinated) or down (not yet pollinated)? • Look closely at the stem to see scale-like leaves.

For your sketches, notes, observations

POISONOUS	**Indian Poke or False Hellebore**
	Veratrum viride
About the names	• <u>Indian Poke</u>: early settlers found Indians using this for various purposes; its early leaves resemble those of Pokeweed. • <u>Hellebore</u> means "poison-food," which this is, so why it is styled <u>False</u> is a mystery. • <u>Veratrum</u> means "truly black"; <u>viride</u> means "fresh-green" (these names may refer to its direly poisonous nature but pretty green color). • Also called Green or White Hellebore, Bear-Corn, Bugwort, Crow-Poison, Devil's-Bite.
Memory aid	• Its clasping leaves are heavily ribbed, looking like the <u>hell</u>ish skeleton you'd be if you ate it, your ribs <u>poke</u>-ing out.
When in bloom	• Shoots as early as February, leafs out before deciduous trees; flowers April/May-July.
Interesting growth habits	• Found in wet, swampy areas, not usually right in the water. • Flowers are star-shaped, yellow-green; grow in big branching clusters at top of plant. • Can grow to 6 feet tall, with leaves up to a foot long and 6 inches wide.
Warnings, uses	• CAUTION: All parts of the plant, but especially the root, contain highly toxic alkaloids which seriously affect the heart and nervous system. • Has been used as a heart treatment (sometimes with drastic results). • Used as a poison in ancient times.
"Stories"	• "The Indians cure their wounds with it, annointing the wound first with raccoons greese, or wild-cats greese, and strewing upon it the powder of the roots...the powder of the root put into a hollow tooth, is good for the tooth-ache." (1672) • "The fresh roots, beaten up with hog's lard, cures the itch...crows may be destroyed by boiling Indian corn in a strong concoction of the fresh roots, and strewing it on the ground where they resort." (1785)
Do this	• Find some Skunk Cabbage (q.v.)and compare the two. • Measure the biggest leaf you can find.

For your sketches, notes, observations

POISONOUS	**Jack-in-the-Pulpit**
	Arisaema triphyllum
About the names	• <u>Jack-in-the-Pulpit:</u> perfectly descriptive. • <u>Arisaema</u> means "Arum blood," Arum being the family name; <u>triphyllum</u> refers to the leaves, each of which is divided into three parts. • Also called Indian Turnip, Wake-Robin, Marsh-Pepper, Starchwort.
Memory aid	• "Jack" standing up in his pulpit, with its little roof sheltering him, is unmistakable.
When in bloom	• April-June/July; fruit July-on.
Interesting growth habits	• Jack is the spadix (with tiny flowers at its base); his pulpit is the spathe. • Each plant must "decide" every year if it will be male or female; a large plant is usually female because it must make enough energy to produce fruit and seeds. • If growing conditions are not good, plant will be male and smaller. • In spring, a fungus-like smell attracts insects, which may then drown in the base of the spathe.
Warnings, uses	• CAUTION Contains calcium oxalate crystals, which cause a violent burning sensation if eaten. • CAUTION Touching roots causes severe irritation. • Root, if and only if thoroughly dried, used as tea for swelling of snake bites, treatment for asthma, headaches; many other traditional uses.
"Stories"	• Roots used to be used to make starch, but it was "most hurtful to the hands of the laundress that hath the handling of it" (1633), and fashionable gentlemen complained of the neck rash caused by their starched ruffs. • It is said that some Native Americans ground up the root, mixed it with food, and presented this to their enemies—who then died horribly.
Do this	• Look inside plant to see its sex: female has tiny green "berries" clustered around base of column, male has thread-like pollen structures. • Look gently in the cup to see if any insects have met their end within. • In late summer and fall, look for clusters of large scarlet berries on a bare stalk.

For your sketches, notes, observations

	Jewelweed or Touch-Me-Not *Impatiens capensis*
About the names	• <u>Jewelweed</u>: the bright orange spotted flower is certainly jewel-like; <u>Touch-Me-Not</u>: don't touch the seedpods in fall or they will explode! • <u>Impatiens</u> means "impatient," to describe its touch-sensitive nature; <u>capensis</u> means "of the Cape," meaning the Cape of Cape Town, South Africa, home to many in this genus. • Also called Horns-Of-Plenty, Kicking-Horse, Jack-Jump-Up-And-Kiss-Me, Snapweed.
Memory aid	• Beautiful speckly vividly-colored jewel.
When in bloom	• July-October or first frost; fruits October.
Interesting growth habits	• Nectar about 40% sugar, attracting bees and hummingbirds. • An annual, growing each year from last year's seed; may be as many as 1000 plants/sq. meter. • In order to ensure cross-pollination, male part matures and declines before female part matures. • Along with bright orange flowers, makes flowers that never open, and pollinate themselves (cleistogamous flowers); more of this type are produced in habitats that aren't optimal.
Warnings, uses	• Effective (truly) on Poison Ivy rash and other itches, such as mosquito bites (grab a handful of the plant, crush it, and rub the mass over the itchy place); a study showed this effective in 108 out of 115 patients over 2-3 days.
"Stories"	• Insects which can't reach nectar by entering the plant from the front may cut holes in the rear tube and steal nectar without taking away pollen (see if you can find any holes).
Do this	• Pick a leaf, immerse it in water, admire its silvery sheen; when you pull it out, it's dry; use a magnifier to see if you can see the tiny hairs that trap air and produce this effect. • To explode the seed pods, look for a fairly fat one (light green); gently encircle it with your fingers and feel it explode in your hand! Then look to see the seed and the springs that propelled it. Open the seed; it has another surprise, a lovely turquoise color. • Look in leaf axils for tiny green pendant cleistogamous flowers.

For your sketches, notes, observations

	Joe Pye Weeds
	Eupatorium species
About the names	• <u>Joe Pye</u>: possibly an Indian healer who taught settlers to use this plant to cure fever and typhus. • <u>Eupatorium</u> is for King Eupator of Parthia (120-63 B.C.), a great opponent of Rome, who is supposed to have discovered the medicinal value of this genus.
Memory aid	• Visualize: Joe Pye standing tall and sturdy, with his crown of fuzzy hair. • Or: the fragrant odor of a delicious "pie" brings insects to investigate.
When in bloom	• July-September.
Interesting growth habits	• When it blooms, fall is coming! • Can reach ten feet tall (though not usually). • Leaves arranged in offset whorls around the stem, so as to get the maximum amount of sun. • A member of the Composite Family, which usually has both ray and disk flowers (like Daisies), but this one has only disk flowers.
Warnings, uses	• Spicebush Swallowtails and other butterflies, as well as bees, feed on these flowers. • Used as diuretic; for kidney problems including stones; for gout;, impotence; bed-wetting—and chronic coughs, among others.
"Stories"	• One source says that Joe Pye was not an Indian at all, but a Caucasian who promoted Indian lore in the 1800's. • Manasseh Cutler, a Yale graduate of the 18th c., a man of many and varied achievements, including a book on New England's plants, reported he'd been told that a dose of Joe-Pye Weed "vomits and purges smartly."
Do this	• Watch a stand of Joe Pye flowers for a while to see what butterflies and other visitors come to sample the nectar. • Smell them yourself, but be cautious so as not to disturb a bee. • Crush some leaves; if they smell like vanilla, you have a Sweet Joe Pye flower (as opposed to some of the other species in the genus). • Check to see the offset whorls of leaves.

For your sketches, notes, observations

CAUTION	**Lady's Slipper**
	Cypripedium acaule

About the names	• <u>Lady's Slipper</u> describes the flower shape well. • <u>Cypripedium</u>: "Aphrodite's foot" (Kypris a name for Aphrodite); <u>acaule</u> means "without an obvious stem," as the flower grows directly out of the stem. • Also often called Moccasin-Flower or American Valerian.
Memory aid	• Along with Jack-in-the-Pulpit, probably one flower most everybody can name.
When in bloom	• May-June.
Interesting growth habits	• It's an orchid! One of the largest families of flowering plants. • Roots of a newly-germinated plant MUST hook up with underground fungi in order to obtain nourishment. • May take several years to come into bloom. • Cannot be transplanted, as one would need to transplant all the tangled roots and connecting fungi with it. • Pollinators must push their way between the lips of the pouch, and can only exit through holes in the back, near which are the anthers with the pollen, as well as the stigma on which pollen from another flower must be left. • Watch out, pollinators! The way in is complex, there isn't much nectar, and there may be predatory spiders waiting inside to catch you! • Seeds are as small as dust, among the smallest.
Warnings, uses	• CAUTION Handling may cause rash. • Victorians used it for "hysteria," depression caused by sexual abuse, and PMS; the suggestive shape of the flowers led to its use as an aphrodisiac.
"Stories"	• Few are actually pollinated; if pollinated, flower will fall off and ovary below will develop into a ridged pod; if not pollinated, only a little green flag will remain on the end of the stalk.
Do this	• On your walk, count to see how many have been pollinated and how many haven't; if a pod has split, it means the seeds have been released.

146

For your sketches, notes, observations

	Wild Lettuces
	Lactuca species
About the names	• <u>Lettuce</u> comes from the root word for "milky" (referring to the sap). • <u>Lactuca</u> comes from the same root and also means "milky."
Memory aid	• The plant is very tall and straight, up to ten feet high: visualize "Let-us stand tall and eat our lettuce."
When in bloom	• July-September.
Interesting growth habits	• Leaves are highly variable in form. • Flowers like small Dandelions, but some species have white, some yellow, and some blue flowers, always numerous, described as "insignifcant." • Milky sap is sticky, like latex. • Seeds look like Dandelion seeds, with feathery "parachute."
Warnings, uses	• Cultivated lettuce is in the same family; young Wild Lettuce leaves can also be eaten raw or cooked. • It "abateth bodily lust, outwardly applied to the testicles with a little camphire." (1652) • The dried milky sap looks and smells like opium and may even have some sedative qualities. • "Lettuce, by reason of its Soperiferous quality...allays heat, bridles choler, extinguishes thirst, excites appetite, kindly nourishes, and above all represses vapours, conciliates sleep, mitigates pain; besides the effect it has upon the morals, temperance, and chastity." (1699)
"Stories"	• Apparently Lettuce was so highly efficacious that "the great [Roman Emperor] Augustus, attributing his recovery of a dangerous sickness to [it]...erected a statue, and built an altar to this noble plant." (1699)
Do this	• Gently break a stalk or leaf and test the stickiness of the sap; sniff it to see its odor.

For your sketches, notes, observations

	Purple Loosestrife
	Lythrum salicaria
About the names	• Loosestrife is a mistranslation of a Greek name that was taken to mean "ending strife," espe cially between farm animals yoked together.
	• Lythrum: "black blood" (the flower color); salicaria: "willow-like," possibly since it grows in wetlands like some Willows, and its leaves are somewhat willow-like.
Memory aid	• Big purple plants crowded tightly together: visualize the strife among them, which would loosen if they were farther apart.
When in bloom	• June-September.
Interesting growth habits	• Three forms of flowers, one form to a plant, each with a different height of male and female parts; Darwin studied these and found that like best fertilizes like.
	• Single plant may make 3000 flowers (and up to 2 million seeds!) in a season.
	• An aggressive plant, it spreads by seeds, by shoots, and even from rooting stem fragments.
	• Main stem has 6 sides, side branch es have 4 sides and grow in whorls of 3; flower clusters come in 2's, with 3 flowers each.
Warnings, uses	• Compound found in leaves stops bleeding and is antibacterial.
"Stories"	• Arrived here about 150 years ago, probably in ship ballast or as a garden pl ant.
	• Although classified in many states as a "noxious weed," which chokes out native vegetation, it is still sold by nurseries.
	• No bird, fish, or mammal is known to eat it; it has crowded out native species which used to provide food and shelter for birds and mammals.
	• Hang a wreath of this on your oxen's yoke, and they will plow in harmony.
	• Experiments for controlling the plant are under way; two species of imported beetles which eat stems, leaves, and buds are showing some promise. (BEWARE...)
Do this	• Use magnifier to look for three flower forms, with stamens and pistils of different lengths.
	• Try finding some of the 2's, 3's, 4's and 6's mentioned above.

For your sketches, notes, observations

Red Maple
Acer rubrum

About the names	• Red Maple: a part of the tree, for which "maple" is the Old English name, is red in all seasons.
	• <u>Acer</u> means "sharp," either because its wood was used for lances, or from the shape of the leaves (related to "oak" and "acre"); <u>rubrum</u>: "red."
	• Also called Scarlet Maple, Swamp Maple.
Memory aid	• No other tree has so much red on it in all seasons: flowers, leaf stalks, twigs, fall foliage, buds.
When in bloom	• February/March-May; fruits March-July.
Interesting growth habits	• Likes wet feet and found in swamps, but also found in nearly every other forest type.
	• A single sex of flowers often found on one branch or even one tree; male flowers tend to be more orange-red than female, which are pure red.
	• Sprouts easily from stumps and can crowd out other species.
	• Seeds (called keys or samaras) in pairs with tiny "wings," attached at base for a helicoptering function as they fall to earth, the weightier seeds hitting the soil first.
	• The redness of the fall foliage depends on acidity of soil: the more acid the redder.
	• Bark on old trees may seem to be peeling off in long, up-curling scales.
Warnings, uses	• Early settlers made inks and dyes from the bark.
	• Mammals and ground-walking birds browse on the twigs.
	• Numerous kinds of caterpillars eat the foliage.
"Stories"	• Maple syrup can be made from the sap of this tree, although it's even more dilute than that of the Sugar Maple.
	• Do squirrels seem to be eating the flowers on your Maple? They are actually eating the young seeds.
Do this	• Use a magnifier to enjoy the delicate red flowers in spring; male flowers have 5-8 long stamens, female flower parts are shorter.

For your sketches, notes, observations

	Striped Maple
	Acer pennsylvanicum
About the names	• <u>Striped</u> refers to the bark; <u>Maple</u> is an Old English word for this tree. • <u>Acer</u>: "sharp," Latin, referring possibly to its leaves or its ancient use in making lances; <u>pennsylvanicum</u>: "from Pennsylvania." • Also called Moosewood.
Memory aid	• Distinctive white and green striped bark is unmistakable.
When in bloom	• May-June; fruits in June-September.
Interesting growth habits	• Beautiful green bark with vertical white stripes. • Plain-looking three-lobed leaves can reach eight inches wide or more. • Winged "helicopter" seeds are pitted on one side. • Trunk may grow no more than eight inches wide.
Warnings, uses	• A report in 1885 told of Native Americans' use for this tree: "The inner bark scraped from four sticks or branches, each two feet long, is put into a cloth and boiled, the liquid which can subsequently be pressed out of the bag is swallowed, to act as an emetic." • Also used as an anti-inflammatory agent, and a similar species from Asia has been shown to contain a compound which indeed is medically effective for this purpose.
"Stories"	• A favorite food of moose, who have to eat, per day, about four pounds of leaves for every eighty pounds of their weight; that is a lot of leaves! • Some Native Americans believed that a bad spirit resided in the Striped Maple leaves, and would not have one near their homes. • More lore: Do not cut this tree for firewood, or you will starve to death!
Do this	• During May, look for the lovely pendant strings of chartreuse flowers. • During summer, what's the biggest Striped Maple leaf you can find? • In late summer, find a seed pod and look for the little pit on one side.

For your sketches, notes, observations

	Sugar Maple
	Acer saccharum
About the names	• <u>Sugar</u>: the tree with sweet sap; <u>Maple</u> appears to be a unique Old English word, particular to this tree. • <u>Acer</u> means "sharp," either for the shape of the leaves or because the wood was used for lances; <u>saccharum</u> means "sweet" or "sugary." • Also called Rock Maple or Hard Maple.
Memory aid	• Look at the spaces between the lobes of the leaves; they are U-shaped, like the U in sUgar. • In winter, see brown buds on brown twigs: brown like maple syrup.
When in bloom	• April-June, along with the leaves; fruits June-September.
Interesting growth habits	• Can live to three hundred years. • There may be fifty male flowers to one female on a tree. • Leaves contain high percentage of calcium, which maintains limy soil underneath, congenial to earthworms and to the tree itself.
Warnings, uses	• Native Americans taught Colonists about syrup- and sugar-making. • Wood is used for fine furniture and for musical instruments; also for bowling alleys and pins. • What would New England be without its fragrant sugarhouses in spring and gorgeous Sugar Maples in fall, glowing yellow, orange, and scarlet!
"Stories"	• Trees must be 40-80 years old before they yield syrup-making sap; each tree gives between 5 and 60 gallons of sap a year, and it takes about 30 gallons to make one gallon of syrup, but syrup-making sap can only be collected for about 3 weeks in spring, when days are warm and nights are cold. • Squirrels like the sweet sap too; they may gnaw the bark and lick the sap that then drips out.
Do this	• In early spring, try just nicking a twig to see if sap drips out; if you taste it, you will be disappointed, since most of it is water. • Look for the double-winged fruits and play with them to see how they twirl to the ground. What happens if you hold them in different ways before letting them go?

For your sketches, notes, observations

POISONOUS	**Marsh Marigold**
	Caltha palustris
About the names	• <u>Marsh Marigold</u>: describes habitat and color. • <u>Caltha</u>: "cup"; <u>palustris</u>: "from boggy or wet ground." • Also called Cowslip.
Memory aid	• Found in marshes, the color of Marigolds.
When in bloom	• April-June.
Interesting growth habits	• Its shiny yellow flowers look like big Buttercups, and they are actually in the Buttercup family. • Stem is hollow and grooved on one side. • Those are sepals, not petals; flower actually has no petals. • Seed pods look like brown flowers themselves. • Leaves grow bigger during the summer, to produce food to power next spring's flowering.
Warnings, uses	• CAUTION All parts of the plant are toxic. • CAUTION May cause skin irritation. • Native Americans mixed its tea with maple sugar for a cough medicine. • A syrup made from it was supposed to be effective against snakebite. • Flowers used for dye. • Leaves can be eaten, but must be lengthily boiled in several changes of water to rid them of poison.
"Stories"	• May, when it is in bloom, was in olden times the month in which the Virgin Mary was celebrated; perhaps the golden flowers were used as decoration in ceremonies honoring her, hence "Marigold." • This is one of those flowers which, in the ultraviolet wavelength in which bees see, looks like a brilliant target with "honey-guide" lines leading to a dark center, wherein are the nectar and pollen. • "And the wild marsh-marigold shines like fire in swamps and hollows gray," wrote Tennyson.
Do this	• Can you get close enough to see the grooved stem? • Try scraping off the petals' yellow color! • Post-flowering, see if you can find seed pods.

For your sketches, notes, observations

	Canada Mayflower
	Maianthemum canadense
About the names	• Possibly originally found in <u>Canada</u>, and blooms in <u>May</u>; the Latin name means the same.
	• Aalso called Cowslip, Wild Lily-of-the-Valley, Bead Ruby, Elf-Feather (this one very descriptive).
Memory aid	• <u>C</u>ommon <u>C</u>anada May flower (<u>very</u> common!).
When in bloom	• May-June; fruits July-August.
Interesting growth habits	• Usually just two leaves per stem; heart-shaped at their bases.
	• But some plants have only one leaf and no flowers: if the plant doesn't have enough leaves, it can't produce enough energy to flower.
	• Flowers, in small clusters, are tiny star-bursts.
	• May form extensive light green swathes in the spring woods, sometimes only 3 inches high, spreading by underground runners.
	• Its presence indicates a strongly acidic soil.
Warnings, uses	• Berries are edible but not very tasty.
	• Root was used as a good-luck charm in games.
"Stories"	• Big mats of this plant don't appear in younger woods or relatively recently abandoned farmlands, because early pioneer species crowd it out; so its presence indicates a mature forest.
Do this	• Check to see if one-leaved plants do or don't have flowers.
	• Use a magnifier to see the delicate flower parts.

For your sketches, notes, observations

	Meadowsweet
	Spiraea latifolia
About the names	• <u>Meadowsweet</u>: probably a corruption of an older name, Meadsweet, from its oldest name, Meadwort, meaning a plant used to flavor the drink mead. • <u>Spiraea</u>: name given by Theophrastus (307-285 B. C.) to plants used for making garlands; <u>latifolia</u> means "broad or wide leaved."
Memory aid	• Not many flowering bushes in the meadow: it's <u>sweet</u> to find a flower-bush in the <u>meadow</u>.
When in bloom	• June-September.
Interesting growth habits	• Clusters of pinkish-white flowers with long protruding stamens grow in a kind of pyramid shape; top flowers are male and bloom first, lower ones are female. • In fall and winter, the dry seed pods look like clusters of lacy brown flowers. • Twigs are reddish or purplish-brown.
Warnings, uses	• Contains salicylates, a component of aspirin; used as a remedy for flu, fever, and arthritis. • A good tea can be made from the leaves.
"Stories"	• In Newfoundland, it's called Dead Man's Flower, and if you pick it, your father will die. • When bees land on the plant, they go first to the female flowers at the bottom, then work their way up to the male flowers, so when they go to the next plant's female flowers, they are carrying pollen from another plant's male flowers. • In the Shetland Islands it's called Courtship and Matrimony Plant, since the smell changes once the leaves are walked on (they were used on floors). Not a pretty story!
Do this	• Look closely on the flower for the pink and white Goldenrod spider, and for other insects. • Look underneath the leaves: if they are pale brown and fuzzy, it's another Spiraea, Steeplebush or Hardhack, with pinkish flowers. • In winter, look for a spiny growth on top a stem, a gall caused by a little fly laying her eggs in the stem. • Count the number of branchings under the flower head to find the age of the plant.

For your sketches, notes, observations

POISONOUS	**Milkweed**
	Asclepias syriaca
About the names	• <u>Milkweed</u> refers to the milky sap. • <u>Asclepias</u>: for Asclepius, Greek god of medicine, Apollo's son; <u>syriaca</u> means "from Syria" (named by Linnaeus, the plant was introduced to Europe from America; he thought it came from Syria).
Memory aid	• Drooping balls of flowers and milky sap remind one of milking heavy udders.
When in bloom	• June-August.
Interesting growth habits	• Stamens and pistil are on a club, hidden behind a hood; only some bees and butterflies are strong enough to pull this down to get at nectar and pollen; pollen is in clumps (called pollinia), and insects must be strong enough to detach these. • Seeds released only a few at a time; if all released at once they would stick together and compete with each other for ground space.
Warnings, uses	• CAUTION Contains toxic substances. • Native Americans used for a number of treatments: kidney stones, warts, ringworm. • "Silk" used to stuff pillows, and during World War II, lifejackets.
"Stories"	• Monarch butterflies lay their eggs on Milkweed in spring; when larvae hatch and eat the plant, they absorb its toxins, thus rendering both caterpillar and butterfly sickening to predators (a famous film in zoology is of a young blue jay eating a Monarch caterpillar for the first time and spitting it up!). • The abundance of insects attracted to Milkweed in turn attracts insect-eaters such as spiders. • The down "may be spun into...excellent wickyarn [better than that] made with cotton...they will not require so frequent snuffing and the smoke of the snuff is less offensive." (1785)
Do this	• How many different insects and butterflies are found at a patch of Milkweed in 10 minutes? • With a magnifier, examine the complex flower. • Being careful not to inhale an insect, smell the sweet scent. • In fall, look for the seed pods with their wondrous huge silky parachutes.

For your sketches, notes, observations

	Hair-Cap Mosses *Polytrichum* species
About the names	• <u>Hair-Cap</u>: refers to the "hairy" covering on the spore capsule covering. • <u>Polytricum</u> means "many hairs." • Also called Pigeon Wheat or Goldilocks.
Memory aid	• A mat of this moss, with its spore capsules standing up on their thin filaments, does indeed look hairy (although the name does not refer to this part of the plant).
When in fruit	• Spores produced in warm months.
Interesting growth habits	• Green parts are either male or female; male plants produce a flower-like structure, which fertilizes female part; female part then grows a tall filament, at the end of which grows the spore capsule; "successful" spores eventually germinate new green parts. • Note that the green "leafy" part of the moss and the stalked spore capsules are two different generations of the same plant; the spore capsules are dependent on the "leafy" parts for nutrients. • Spore capsules of Hair-Cap Moss are larger than those of similar mosses. • Mosses absorb water and nutrients directly through their cells; they have no internal conducting system as do larger plants.
Warnings, uses	• Used to treat kidney stones and as a diuretic. • Used for mattress stuffing.
"Stories"	• Over 100 species of Polytrichum mosses. • The hairy cap has a lovely name; it's called a calyptra.
Do this	• Use a magnifier to examine the elegant spore capsule—is the hair-cap still on it? • If the hair-cap is gone, see the white membrane covering the opening of the capsule.

For your sketches, notes, observations

Sphagnum Mosses
Sphagnum species

About the names	• <u>Sphagnum</u>: from a Greek word, via Latin, for a kind of shrub; further meaning not known. • Also called Peat Moss.
Memory aid	• "SPhagnum SPonge" for its ability to hold water.
When in fruit	• Spore capsules form in spring.
Interesting growth habits	• Makes bogs acidic by exchanging mineral ions from water for hydrogen ions. • Absorbs nutrients directly through surface cells, rather than through a vascular system, of which they have none. • Also has no true roots, stems, or leaves. • Reproduces by spores rather than seeds. • It grows at the top and dies at the bottom, thus forming thick, thick layers below (peat), which release acids and tannins, thus preventing decay. It eventually carbonizes in to soft (bituminous) coal, and if enough time passes for this to metamorphose, it becomes hard coal (anthracite).
Warnings, uses	• Because Sphagnum does not decay (its habitat is too acidic to support bacteria), it has been used as an antiseptic wound dress ing. • Water-retention quality led to its use as diapers by Native Americans. • Peat used as fuel since before Roman times.
"Stories"	• Can hold up to twenty times its weight in water. • In ancient times, bogs were sacred places, and human sacrifices appear to ha ve been made in them; "Bog Men," preserved from Neolithic times, have been discovered, beautifully preserved, in peat bogs. • Absence of decay bacteria has also preserved prehistoric pollen in peat bogs, enabling paleobotanists to understand ancient biomes. • Two tons of peat can produce as much electric power as one ton of coal or four tons of wood. • But a one-inch seam of coal equals centuries of peat formation!
Do this	• Grab a handful (be sure to return it —it will be fine) and squeeze to see how much water it holds.

For your sketches, notes, observations

CAUTION	**Mullein**
	Verbascum thapsus
About the names	• <u>Mullein</u>: from an ancient French word meaning "soft," in turn derived from Latin. • <u>Verbascum</u>: name given by Pliny (23-79 A.D., Roman naturalist, scholar); <u>thapsus</u>: "from Thapsos" (an ancient city in present Tunisia). • Also called Flannel-Leaf, Velvet-Plant.
Memory aid	• Mmmmm...that's soft...mmmmmullein! (lots of words that relate to the concept "soft" come from the same root: mollify, mild, melt, emollient).
When in bloom	• May-September.
Interesting growth habits	• Spends first year as an evergreen rosette. • Basal leaves are very long (up to a foot), thick, soft as flannel, or maybe rabbit fur. • Native to Eurasia.
Warnings, uses	• CAUTION Contains harmful compounds; also may cause skin irritation. • In olden times, winter seed head (to three feet long) was dipped in melted tallow, filling the hundreds of empty seed pods, and set alight as a torch. • Native Americans put its soft leaves in their moccasins, a sort of early Dr. Scholl's "pillo-insoles." • Seeds were used to stun fish as they swam. • Roman ladies dyed their hair with the flowers.
"Stories"	• In 1879, in an experiment, 20 bottles of 100 seeds of 20 "weed" species were buried; in 2000 22 Mullein seeds germinated from bottle #15. • BUT: In Denmark, Mullein plants sprouted from dirt collected from under the foundation of a 650 year old church! • "...some think that this herbe being but carryed... doth help the falling sickness, especially ...gathered when the sun is in Virgo and the moon in Aries, which thing notwithstanding is vaine and superstitious." (1597)
Do this	• Carefully feel the velvety soft leaves (if you have a very strong magnifier, examine the hairs to see the "stars" at their tips). • If it's gone to seed, gently shake the top to see the tiny (6.-.9 mm) black seeds; look closely to see the two-part seed capsules typical of its Snapdragon family.

For your sketches, notes, observations

CAUTION	**Mustards**
	Brassica species
About the names	• <u>Mustard</u>: comes from Latin words meaning "fresh" (used to describe new wine, with which ground Mustard seed was mixed to prepare the condiment), and "burning" (for the hot taste).
	• <u>Brassica</u>: "cabbage," a member of this genus.
Memory aid	• Wow! those seed pods are running up the stem to escape the hot mustard!
When in bloom	• May/June-October; fruits forming throughout.
Interesting growth habits	• Slender, beaked seed pods point up the stem, like tall coat hooks.
	• Flowers' four petals make a cross, hence "Cruciferous" plant family.
Warnings, uses	• CAUTION Mustard oil is a powerful irritant ("among the most powerful caustic agents known").
	• Ground Black Mustard seeds make the powdered kind; a mixture of Black and White, the spreadable (going back to Roman times).
	• Used in salads, as cooked green, pickle-making.
	• Unopened flower buds can be used like broccoli.
	• Ground up seeds once used as a snuff for headaches.
	• Used for "mustard plasters," poultice made of powdered Mustard seed, flour and water, spread between pieces of cloth and placed on chest of cold sufferers.
"Stories"	• Plant rich in vitamins A, B, and C; flowerbuds rich in protein.
	• Same genus as broccoli, cabbage, cauliflower, kale, collards, kohlrabi, turnips, rutabaga.
	• "The powdered seeds, with crumbs of bread and vinegar, are made into cataplasms, and applied to the soles of the feet in fevers, when stimulants are necessary." (1785)
	• "The oil is very little affected by frost or the atmosphere, and is therefore especially prized by clock-makers and makers of instruments of precision...[it also] promotes the growth of the hair..." (1931)
Do this	• Open a seed pod to see the tiny black seeds (Black Mustard) or white ones (White Mustard).

For your sketches, notes, observations

	Nannyberry
	Viburnum lentago
About the names	• <u>Nannyberry</u>: "nanny" meaning goat? If so, may refer to unpleasant, skunky odor of its bark and twigs. • <u>Viburnum</u>: Latin name for Wayfaring-Man's-Tree, a member of this genus; <u>lentago</u> meaning "flexible" (the side twigs are). • Also called Sweet Viburnum (for the fruit), Sheep-Berry (doubtless for the odor).
Memory aid	• If it's got opposite branching and its bark smells skunky, like a nanny-goat, it's Nannyberry.
When in bloom	• May-June; fruits August-September.
Interesting growth habits	• Fruit passes through three colors: yellow to pink to dark blue with a whitish bloom; hangs on drooping reddish stalks. • May grow as big as a small tree (25 feet). • Leaves turn purplish-red in fall. • Long reddish hairy terminal bud looks grainy. • Big flat clusters of tiny white flowers. • Leaf stalks are winged.
Warnings, uses	• A tasty winter fruit for Native Americans. • Fruit also eaten by birds and small mammals. • Jam can be made of fruits.
"Stories"	• May sprout directly from branches bending down and rooting where they touch ground • Although fruits are sweet, they have a low fat content and so are not especially nutritious.
Do this	• Test for that skunky nanny-goat smell. • Look with a magnifier at the underside of leaves for tiny black dots of glands.

For your sketches, notes, observations

CAUTION	**Black (or Red) Oak family**
	Quercus velutina and others

About the names	• <u>Black Oak</u> may refer to the blackish bark. • <u>Quercus</u>: the old Latin name for Oak; <u>velutina</u> means "soft, with a velvety covering," referring to the hairy buds, as well as the acorn cups, and the twigs, which may be downy.
Memory aid	• Leaves in this family all have bristled tips to their lobes, sticking out like the ends of the letter K in <u>black</u>; buds are pointy too.
When in bloom	• May-June.
Interesting growth habits	• Inner bark is orange or yellow. • Acorns take two years to mature, so there may be both one and two year old acorns on one tree. • Male flowers look like hanging strings of beads; female are greenish, yellowish, or reddish, found (a few) in axils of new leaves. • Produces a big crop of acorns every 3-4 years.
Warnings, uses	• CAUTION Tannin is potentially toxic. • Inner bark used for dye and as a source of tannin for tanning leather; an Oak leaf can contain 60% tannin! • If acorns are boiled in enough changes of water, and then dried and ground, they can be made into nutritious flour. • Like White Oak, tannic acid from this tree has been shown to be antiviral, antiseptic, anticancer—and carcinogenic!
"Stories"	• Black Oak family acorns have both more fat and more tannin, and sprout later, than White Oak acorns; squirrels want the fat but not the tannin, so they tend to bury these seeds rather than eat them right away (as they do White Oak acorns), apparently saving them for times of need.
Do this	• Look for different ages of acorns on one tree. • See if you can find a White Oak leaf to compare. • Look at the overall shape of the tree: did it grow in the open, making a short, thick trunk and wide crown, or in the shade, consequently being tall and straight with a narrow crown? • See how many galls you can find on an Oak tree: big round oak apple galls, clusters like grapes, and many others—over 300 kinds!

For your sketches, notes, observations

CAUTION	**White Oak family**
	Quercus alba and others

About the names	• <u>White Oak</u>: possibly for whitish bark; also leaves are whitish below, nearly white when new. • <u>Quercus</u>: old Latin name for Oak tree; <u>alba</u> means "white."
Memory aid	• Lobes of leaves are rounded, curving, soft: visualize similar <u>white</u> things such as clouds, cotton, melting vanilla ice cream (note also that clustered buds are rounded).
When in bloom	• May-June.
Interesting growth habits	• Acorns mature in one season; the warty cap covers only about a third of the nut. • In a good year, one tree may make 2-7000 acorns. • Male flowers like clusters of hanging beads; the few female flowers, found in the axils of new leaves, are greenish or reddish. • Leaves may stay on trees all winter. • Many different insects lay eggs within White Oak, producing several hundred different types of galls (excrescences formed by tree's tissues in response to chemicals produced by larvae).
Warnings, uses	• CAUTION Tannic acid is potentially toxic. • Tea made from inner bark used for sore throats, hemorrhoids, dysentery, poison ivy, cancer. • Acorns are described as "sweet" (here a relative term); after being hulled and boiled repeatedly, can be roasted and eaten plain or dipped in sugar syrup, or ground up for flour. • Acorns are important food for numerous animals.
"Stories"	• Tannic acid found to be antiviral, antiseptic, antitumor—and carcinogenic. • "The outstanding wood for tight barrels" (sometimes called Stave Oak) • White Oak was used for the gun deck of USS Constitution (Old Ironsides). • Squirrels and blue jays bury the nuts and then forget them, thus "planting" new trees. • Turkeys eat the acorns whole; one account says up to 50 in one meal!
Do this	• Find an acorn to examine for the small warty cap, and take it and a leaf to carry with you to compare with other Oaks.

For your sketches, notes, observations

	Partridgeberry
	Mitchella repens
About the names	• <u>Partridgeberry</u>: supposedly the fruits are eaten by partridges.
	• <u>Mitchella</u> honors John Mitchell, who lived in Virginia after 1725; his maps of the colonies became famous, but he also was interested in the plants he found in his travels, and corresponded with Linnaeus about them; <u>repens</u> means "creeping."
	• Also called Squaw Vine, Checkerberry, Deerberry.
Memory aid	• Visualize: partridges, ground-dwelling birds, walking along the forest floor, spying this low creeping plant, and eating up its bright berries.
When in bloom	• June-July; fruit August-September and winter.
Interesting growth habits	• Twinned flowers have fused ovaries; the red berry shows two spots where each flower was attached to the ovary.
	• Flowers are either male (full stamens and reduced pistils) or female (vice versa).
	• Evergreen; indicates acid soil.
	• Low fat content of berries makes them resistant to rotting, so they last through the winter.
Warnings, uses	• Fruits can be nibbled or added to salads but though pretty, they are dry, seedy, and tasteless.
	• Used by Native Americans for menstrual disorders, easing childbirth, and sore nipples in nursing.
"Stories"	• In 1723 a Colonial botanist sent some specimens to a colleague in London, and told how "Colonel Brown, a man of great figure here, told me his body was so big of a dropsie [edema] as a sack of malt, but by drinking this tea he soon recovered a very healthy state of body...and mynds to drink it to his dying day. It is the beloved food of partridges, and that I take to be the reason why we have the best partridges in the world."
Do this	• Open a berry to find 8 seeds; look for the two spots where the flowers met the single ovary.
	• Examine and smell the fringed flower; check to see if it is male or female.

For your sketches, notes, observations

Sweet Pepperbush
Clethra alnifolia

About the names	• <u>Sweet Pepperbush</u>: the flowers of the bush are extremely fragrant; dried seed pods look like peppercorns. • <u>Clethra</u>: ancient Greek name for Alder (plant is in the White Alder family, which has only this one genus); <u>alnifolia</u>: "alder-like leaves."
Memory aid	• The little rows of gray peppercorn-like seed pods in winter and spring, and the amazing SWEETNESS of the flowers in summer will always identify this shrub.
When in bloom	• June/July-September; fruits September-October.
Interesting growth habits	• Incredibly fragrant flowers. • Usually found in or at the edge of wetlands. • One of the few shrubs that reproduce by underground runners, so there may be a large group of them all together. • Indicates acidic soil.
Warnings, uses	• A nectar source for many insects. • A treat for the walker!
"Stories"	• Called, in the old days, Sailors' Delight, since sailors could smell the flowers far out to sea and know that land was near. • Functionally, the flower is first male and then female, since the different sexual parts mature and present themselves in that order. Flowers in a spike bloom from the bottom up, so the bottom flowers are functioning as female while the top ones, on another spike, are male. So visiting bees, who start working a new plant at the bottom, carry pollen from top male flowers they just left to bottom females. Very nifty.
Do this	• Carefully, watching out for bees, examine the tiny flowers with a magnifier; there are ten stamens. • If you can wet your hands, then rub a handful of flowers, surprise: "soap"! • Use a magnifier to see the tiny hairs on the twigs.

For your sketches, notes, observations

| | **Pickerelweed** |
	Pontederia cordata
About the names	• <u>Pickerelweed</u>: young pike fish ("pickerels") are supposed to browse among these plants. • <u>Pontederia</u>: Guillo Pontedera (1688-1757), professor of Botany at University of Padua; <u>cordata</u> means "heart-shaped," referring to the leaves. • Also called Moose-Ear.
Memory aid	• <u>Pick</u> a s-<u>Pike</u> of <u>Purple Pickerelweed</u>.
When in bloom	• June or earlier-September or later.
Interesting growth habits	• Likes shallow, muddy water near banks. • Can tolerate brackish water, so may be found in marshes near estuarine rivers. • Flowers come in three forms, each with different lengths of sexual parts, only one form in a colony; like must be fertilized by like (same as for Purple Loosestrife). • Each flower is open for only a day. • After blooming is over, flower spike bends over and deposits seeds in water, where they must have about two months' cold to germinate. • Stems have air-filled chambers for buoyancy.
Warnings, uses	• Seeds can be roasted and eaten out of hand or ground for flour. • Dragonflies and damselflies lay their eggs in stems below water line.
"Stories"	• Izaak Walton (1676) said he had heard that some people thought pickerel fish were actually generated by Pickerelweed, so that it "both breeds and feeds them," but he wasn't sure if such pickerel could reproduce themselves, and would leave the investigation of that question "to the disquisitions of men of more curiosity and leisure than I profess myself to have."
Do this	• If you can get close to a flower, see if you can see the yellow spots at the base of the longest upper lobe. • In late summer, look for clusters of irregular greenish fruit. • Look on leaves for cast-off exoskeletons of dragonfly and damselfly nymphs.

For your sketches, notes, observations

Pitch Pine
Pinus rigida

About the names	• <u>Pitch</u> refers to the abundant resin from this tree; <u>Pine</u> and <u>Pinus</u> both mean "pine," derived from a root word meaning "to swell" or "fat" (referring to the resin; some related words are pineal, pituitary, piñon). • <u>Rigida</u> means "stiff," for its stout, stiff needles.
Memory aid	• Needles in 3's, like the three tines of a <u>pitch</u>fork.
When in bloom	• Evergreen.
Interesting growth habits	• Bundles of needles may sprout directly from trunk. • Cones may remain on tree for several years after opening. • Well-adapted to fire: if foliage burned, will sprout needles directly from trunk, if growing tip is burned, will develop a new one; if trunk burned, will readily sprout from roots or stump; bark is thick and protects from fire. • Cones may remain closed until singed; seeds then fall on burned-over, thus cleared, land, with good ash fertilizer. • However, can germinate on virtually sterile soil.
Warnings, uses	• Colonists used the resin, and made turpentine and tar from it. • Its resinous knots were made into torches by attaching to a pole. • Used as fuel for wood-burning locomotives.
"Stories"	• Its often twisted branches may indicate repeated sproutings after fires. • Conifers have been on earth since before the dinosaurs!
Do this	• Check the bundles of needles to count them; note how the needles are often twisted. • Check for cones: tiny inch-long immature ones of first year, and mature unopened and opened ones.

For your sketches, notes, observations

White Pine
Pinus strobes

About the names	• <u>White</u>: the lumber is a creamy white. <u>Pine</u> and • <u>Pinus</u>: Latin genus name, meaning "to swell" or "fat," referring to the resin; <u>strobus</u>: ancient name for an incense-bearing tree.
Memory aid	• 5 needles in a bundle: 5 letters in W H I T E.
When in bloom	• June (leaving a film of wind-borne yellow pollen everywhere).
Interesting growth habits	• After about 20 years, greenish male flowers in clusters all over tree; pinkish female flowers on tips of new twigs near top of tree; after pollination, cone develops, growing to 5-6 inches by end of second year, when seeds drop. • Needles low in nutrients essential for growth of understory species; little grows under these trees except plants adapted to very acidic soil. • Mature at 200 years; and may live to 300 years (the ones we see are mostly 60-80 years old). • Cannot grow in its own shade.
Warnings, uses	• Colonists and English used straight, tall trunks for ships' masts. • Now used for pulpwood and in construction. • Excellent nest tree for owls, hawks. • Food for many insects, birds and mammals. • Finely chopped young needles used for tea (needles have more vitamin C than lemons).
"Stories"	• King's foresters came through woods, measuring White Pine trunks and marking those of diameter two feet and more, to be used by the British Navy; woe unto you if your floorboards were found to measure more than two feet across, showing you'd been poaching Royal property! • It's said that the original Pine forests were so enormous that a squirrel might spend its whole life aloft. Some writers disagree, however, and say that the Pine forests were never that thick. Still, a wonderful image! • The first flag of the American Revolutionary forces showed this tree.
Do this	• If it's a fairly small tree, count the whorls of branches around the trunk: one whorl for each year of growth. • Lie down under, look up, close eyes, relish sounds of wind, scent of Pine in the sun.

For your sketches, notes, observations

| POISONOUS | **Pipsissewa** |
	Chimaphila umbellata
About the names	• <u>Pipsissewa</u> from an Indian word meaning "to break up," since it was used to treat gall or kidney stones. • <u>Chimaphila</u>: "winter-loving" (an evergreen); <u>umbellata</u> referring to flowers all growing from the same point. • Also called Prince's Pine, Bittersweet, Love -in- Winter, Pine-Tulip.
Memory aid	• Like Spotted Pipsissewa, a <u>pip</u> of a find on the forest floor, especially with its shiny <u>sisse</u> leaves that try to hide.
When in bloom	• June/July-August.
Interesting growth habits	• Evergreen. • Tiers of leaf whorls around the stem. • Grows in little colonies, spreading from underground runners. • Very pretty green and shiny leaves that always look brand-new.
Warnings, uses	• CAUTION Leaves bound on skin may cause blisters. • CAUTION Contains potentially toxic compound. • Used as diuretic, urinary antiseptic, antibacterial. • Native Americans smoked leaves like tobacco. • Chewing a leaf a day is supposed to ward off tuberculosis. • Used as flavoring in a kind of root beer.
"Stories"	• It's said that tasting a leaf will give a flavor astringent, sweet, and bitter at the same time; also that this peculiar taste will take an hour or so to leave your mouth.
Do this	• If you find it in bloom, look for the ring of tiny red or purple anthers within, and smell it for the sweet scent.

For your sketches, notes, observations

CAUTION	**Spotted Pipsissewa**
	Chimaphila maculata
About the names	• <u>Spotted Pipsissewa</u>: a Indian word meaning "to break up," as it was used to treat kidney stones; it is striped rather than spotted. • <u>Chimaphila</u>: "winter love" since it is evergreen; <u>maculata</u> means "spotted" even though it isn't. • Also called Striped or Spotted Wintergreen, Striped Prince's Pine.
Memory aid	• A <u>pip</u> of a find on the forest floor, since it's a <u>sisse</u> and hides there!
When in bloom	• June-August.
Interesting growth habits	• An evergreen. • Spreads both via runners and by seed, especially after wildfires. • Its ground-hugging size, dark color and mottling camouflage it well. • Indicates an acid soil.
Warnings, uses	• CAUTION May irritate skin. • Used as a diuretic.
"Stories"	• "The plant is in high esteem among the natives...I have myself been witness of a successful cure made by a decoction of this plant, in a very severe case of hysteria. It is a plant eminently deserving the attention of physicians." Thus said Frederick Pursh, born 1774 in Saxony. He came to America to manage a garden near Baltimore, made collecting trips all over Eastern U.S., and was chosen to classify the plants collected on the Lewis and Clark expedition. Later he went to Canada to compile a flora there, but all his work was destroyed in a fire and he died destitute in 1820.
Do this	• Lucky you if you find it in flower! If you do, smell it for its sweet fragrance.

For your sketches, notes, observations

POISONOUS	**Poison Ivy**
	Rhus radicans

About the names	• <u>Poison Ivy</u> is indeed poisonous, to touch and to eat, and it sometimes, but not always, grows in vine form.
	• <u>Rhus</u>: from an ancient Greek word for the Sumac, to which this is related (Its forder genus name, <u>Toxicodendron</u>, meant "poison-tree"!) ; <u>radicans</u>: "with rooting stems."
Memory aid	• You can't beat the classic "leaves of three, let it be." Climbing stem is very hairy, with many brown rootlets, the better to grab you with, my dear, heh heh!
When in bloom	• May-July; fruits August-November.
Interesting growth habits	• Leaves may be rather oily-looking.
	• Can appear as a shrub, a ground cover, or a climbing or trailing vine.
	• Leaves are also quite variable in form.
	• Foliage turns bright red in fall, but don't gather a wreath of it for your Thanksgiving table!
Warnings, uses	• CAUTION If leaves are at all bruised and thus opened, the toxic oil will be released. Contact with this oil, even from something else that has touched it (dogs, tools, boots), will very probably result in an extremely unpleasant rash, which can last a long time and may need medical attention. The oil can also be carried by ashes of burning plants. Plants are potent even in winter, so don't grab hold of a tree to steady yourself on an icy trail without first checking to see if it's got those hairy ropes climbing up it!
	• Song birds and game birds feast on the fruits and thus disperse the seeds.
"Stories"	• Botanists have been stricken with the rash from handling specimens over a hundred years old.
Do this	• STAY AWAY FROM IT!
	• In early summer, cautiously inspect to see if you can spot the yellow-green flowers, and in fall, the grayish-white waxy fruits.

For your sketches, notes, observations

POISONOUS	**Pokeweed/Poke**
	Phytolacca americana

About the names	• <u>Poke</u> apparently comes from a Native American word <u>pokan</u>, meaning "bloody," referring to the fruit juice. • <u>Phytolacca</u>: "plant-dye," referring to the fruit; <u>americana</u> since it is a native plant.
Memory aid	• Don't let that tall, dangerous plant poke you in the eye!
When in bloom	• July–September.
Interesting growth habits	• May reach ten feet, which is quite a sight in late summer when the thick stalks turn magenta! • After cold weather and frost, huge stalks lie prostrate across each other, their brilliant color fading. • Seeds must pass through bird/animal digestive tracts in order to germinate.
Warnings, uses	• CAUTION Roots, stems, leaves, seeds and fruit are extremely poisonous, even fatal; children have died from eating the fruits; may even cause chromosomal damage. • Nevertheless, young shoots up to six inches may be boiled for half an hour in two changes of water, then drained and served on toasted whole grain bread with clarified butter—if you dare. • The berries were used by the Colonists to make a dye. • Many fruit-eating birds eat the berries; one source says they may become intoxicated from fermented fruit!
"Stories"	• "It is doubtful the plant will cure syphilis without the help of mercury." Doubtful indeed! • In Portugal, the berries were used as a coloring for port wine, but the practice ended when it was found to impart a bad taste. And possibly stomach upsets.
Do this	• Admire the gigantic and flashy magenta stems and huge glistening arrays of deep purple berries. • Is anyone in your group wearing clothing that matches the stems?

For your sketches, notes, observations

	Yellow Pond-Lily
	Nuphar variegatum
About the names	• <u>Yellow Pond-Lily</u>: well, it's a yellow Lily that grows in a pond. • <u>Nuphar</u>: a Sanskrit word for Blue Lotus; <u>variegatum</u> means "irregularly colored." • Also called Bullhead Lily or Cow-lily; may be called Spatterdock.
Memory aid	• The name is perfectly descriptive.
When in bloom	• May-September.
Interesting growth habits	• Water-Lily stems contain tubes which aid in floating; they also draw gases up from the sediment, which are diffused through tiny openings on the upper surface of mature leaves. In younger leaves, air is drawn into the openings, down the tubes, and into the roots —all of this operating like a pump. • Six deep yellow petal-like sepals rise like a cup around the yellow-green bottle-shaped sexual parts, the stigma like a flat disk on top.
Warnings, uses	• Native Americans gathered the gummy seed pods, dried them, took out the seeds and roasted them to eat like popcorn or grind into flour. • The root, crushed and soaked in milk, is useful against beetles and cockroaches.
"Stories"	• "The Indians eat the roots...which taste like the liver of a sheep. The moose deer feed much upon them, at which time the Indians kill them, when their heads are under water." (1672) (Moose heads under water, not the Indians') • One study found that 22 liters of air moved through one stalk in a single day. • Numerous insects feed on the leaves, each leaving a telltale shape of hole behind.
Do this	• If you can reach a leaf, feel it to experience the waxy coating that prevents water-logging and damage to the leaves from pelting rain. • If you can't get near the plants, use binoculars to see how many distinct and different shapes have been chewed in the leaves by insects.

For your sketches, notes, observations

CAUTION	**Queen Anne's Lace**
	Daucus carota
About the names	• <u>Queen Anne's Lace</u>: possibly because Queen Anne, wife of King James I, used to decorate her hair with it; alternatively, for St. Anne, patron saint of lacemakers. • <u>Daucus</u> is the Latin name for "carrot," the family of which this is a member; <u>carota</u> also means "carrot." • Also called Wild Carrot, Bird's-Nest, Lace-flower.
Memory aid	• Looks like a crown of lace: Queen Anne's lace crown.
When in bloom	• May-October.
Interesting growth habits	• The larger florets on the outside of the flower head are sterile; they may function as insect attractors; the single deep purple floret in the center of some flower heads may also be an attractor. • In late fall, the dried flower head looks very like a bird nest. • First year growth is a low rosette.
Warnings, uses	• CAUTION CAUTION A similar-looking plant is deadly poisonous! • Used as antibacterial, diuretic, and to expel worms; also as a "morning-after" contraceptive. • "Communicates an agreeable flavor to malt liquor."
"Stories"	• The single purple floret sometimes found in the center of the flower head marks where Queen Anne pricked her finger while making lace. • A flowering plant has been produced from a single cell grown in tissue culture. • Other plants in this family are used as spices (as QAL may be): caraway, fennel, coriander, anise, parsley. • Brought to America by Colonists as a food plant.
Do this	• Crush a leaf to smell the carrot family. • In fall, look at the intricate hooked seeds with a magnifier. • In winter, bend close to admire the display of miniature fireworks, frozen in mid-burst.

For your sketches, notes, observations

CAUTION	**Common Ragweed** *Ambrosia artemisiifolia*
About the names	• <u>Ragweed</u>: for the ragged look of the leaves. • <u>Ambrosia</u>: "food of the gods"; <u>artemisiifolia</u> means to have leaves similar to the plant family Artemisia, which has highly fragrant flowers, suggesting food fit for gods.
Memory aid	• RAGgedy-looking plant with stems curving up, the better to zap you with pollen.
When in bloom	• July/August-October.
Interesting growth habits	• Forms greenish male flowers on stem tips and small female flowers in axils of leaves. • A so-called "fugitive species" that moves its adaptable seeds from place to place.
Warnings, uses	• CAUTION Many are extremely allergic to the pollen of this "despised weed." • CAUTION May cause contact dermatitis. • Native Americans rubbed leaves on insect bites and rashes; used astringent tea for various ailments; root tea for menstrual pain, stroke. • The roots of the Giant Ragweed were chewed to calm night fears. • Pollen of Giant Ragweed is harvested to produce anti-allergy medication. • Seeds are rich in oil and stay on the plant during winter, and so are a useful food for birds.
"Stories"	• Nondescript greenish flowers (as opposed to colorful ones for luring animal pollinators) are sure sign of wind pollination; this is the hayfever culprit, releasing its pollen into the air and thence into our noses. • "When it abounds amongst rye or barley, the seeds are thrashed out with the grain, and will give bread, made of it, a bitter and disagreeable taste." (1785) • Livestock will eat it if it is sprayed with herbicide, thus causing secondary poisoning. • "Ragweeds are probably responsible for more human suffering than any other plant group in the U. S." (1994)
Do this	• DO NOT smell! but search for flower culprits. • In winter, identify by purplish or reddish stems and leaves.

For your sketches, notes, observations

| | **Multiflora Rose** |
	Rosa multiflora
About the names	• All the names mean the same thing: rose with many flowers.
Memory aid	• Unmistakable: indeed a rose with many flowers.
When in bloom	• May-June; fruit September-March.
Interesting growth habits	• Pistils (female parts) are long and form a tall column in the center of the flower, extending beyond the petals. • Leaf-like structures at the base of leaf stalks (called stipules) are elaborately fringed. • Seeds remain viable for 10-20 years.
Warnings, uses	• Introduced from Asia in late 1800's first as a rootstock for other Roses, later as erosion control and "living hedge," but spreads rapidly (probably via birds) and displaces native species; very difficult to get rid of. • Rose hips contain about 100 times as much Vitamin C as oranges, weight for weight. • Thickets serve as good locations for bird nests, especially mockingbirds and catbirds.
"Stories"	• Listed in <u>Weeds of the Northeast</u>! • Mice may open the hips and take out the seed containers to gnaw them open and get the seeds within; they may be using abandoned bird nests in the Roses for their own nests, and leave piles of partly-chewed hips in them. • Spread of mockingbirds and cardinals to our area may be in part due to all the food produced by this flower; robins also eat the hips in winter.
Do this	• Open one of the hips, but first guess how many seeds are inside (actually those aren't seeds, but containers for seeds called achenes). • Look at stems to see fringed stipules. • Look around to see if plundered hips are about. • Listen to hear if a mockingbird or catbird is in the vicinity.

For your sketches, notes, observations

CAUTION	**St. Johnswort**
	Hypericum perforatum
About the names	• St. Johnswort because it was said to bloom on June 24, that saint's day (note that "wort" means "plant"). • Hypericum means "above pictures," in this case pictures in a shrine, as the plant was hung above a shrine to repel evil spirits; perforatum refers to the apparent holes in the leaves, actually translucent oil glands.
Memory aid	• St. John (represented by the cheery yellow of the flowers) scares off witches, depression, and AIDS (the miniscule black dots being pushed to the edges of the flowers by St. John).
When in bloom	• June-September.
Interesting growth habits	• The yellow flowers have tiny black spots on their margins. • The flowers have a peculiar scent, like ambergris or pine resin. • Branches are opposite each other, like a little candelabra.
Warnings, uses	• CAUTION Contains compound that may cause skin burns in light-skinned people after exposure to sunlight. • Apparently sedative, antibacterial, anti-inflammatory. • Presently being studied for use against AIDS. • Used to treat depression; efficacy is uncertain.
"Stories"	• On June 24, St. John's Day, hang a bunch over the door to scare off witches. • Actually, it was even more useful, as its presence would reveal that of witches who were even just passing by, it being so hateful to them that the slightest scent would cause them to take to the air. • In addition, it warded off lightning.
Do this	• Use magnifier to see translucent oil glands on leaves, or just hold leaf up to the light. • Look closely at the marvelous flowers with their dainty surprises. • Smell to experience the odd scent. • If in seed, tip over the seed cups to look at the shiny seeds.

For your sketches, notes, observations

| | **Wild Sarsaparilla** |
	Aralia nudicaulis
About the names	• <u>Sarsaparilla</u> is derived from Spanish and Arabic words meaning "little bramble vine"; this plant is actually not truly a Sarsaparilla (which is a South American plant), but resembles it in its uses.
	• <u>Aralia</u> is an Indian or possibly French Canadian name for this genus; <u>nudicaulis</u> means "naked-stemmed," (cf. with a similar plant which has a bristly stem; also referring to smooth flower stalk).
	• Also called Sasafafarilla, Sassafriller, Saxapril (sounds like a new drug).
Memory aid	• Sarsapa<u>rilla</u>-umb<u>rella</u>: the whorl of leaves shades the ball of flowers, which grows on a separate stalk, as if under a leafy umbrella.
When in bloom	• May-July/August; fruits in September.
Interesting growth habits	• The globes of delicate greenish-white flowers hide modestly under the foot-high umbrella of three sets of five leaflets.
	• Spreads by underground runners and makes colonies.
	• New spring leaves are shiny, reddish, and come in threes, just like early Poison Ivy sprouts!
Warnings, uses	• Spicy beverage (root beer!) made from aromatic root; also supposed to relieve "lassitude" and purify the blood.
	• Root also applied to burns, boils, itchy places, rheumatic joints.
	• "Under the name of Salsepareille, the root of this plant is one of the most important ingredients of the folk medicine of Canada." (Fr. Marie-Victorin, 1935)
"Stories"	• Roots are said to have been used by Indians as travel food on war and hunting excursions.
	• Patent medicines of the 19th c. made frequent use of Sarsaparilla.
	• In the Ginseng family—along with English Ivy.
Do this	• Look for the dark blue or purplish-black berry clusters if it is late in the season.

For your sketches, notes, observations

POISONOUS	**Sassafras** *Sassafras albidum*
About the names	• <u>Sassafras</u>: from the Spanish interpretation of the name given to this tree by Native Americans of Florida. • <u>Albidum</u>: whitish. • Also called Cinnamonwood.
Memory aid	• Sassafras: green as grass (the young twigs); three S sounds in the name and three forms of leaves.
When in bloom	• April-June; fruits August-October.
Interesting growth habits	• Leaves have three forms: one is oval with an unbroken margin, one like a mitten with a thumb, and one with three fingers (the ghost!). • Leaves turn many colors in fall, all on the same tree. • Reproduce by cloning from underground roots. • Reaches its northernmost limit in our area.
Warnings, uses	• CAUTION The oil contains a carcinogen. • Native Americans used it for many purposes, especially as a "blood purifier"; folk use of tea as a spring tonic. • Used as a flavoring for root beer, toothpaste, and gum (until found to be carcinogenic!). • Ground up leaves (filé) are a spice for gumbo. • Spicebush Swallowtail butterflies may lay eggs on it; their caterpillars feed on it. • Fruit has high fat content and is taken by migrating birds as a good source of energy.
"Stories"	• For a time it was second only to tobacco as an export from the Colonies to England, where it was in great demand for its medicinal powers. (1754: "Joyful Newes out of the Newe Founde World"—a cure for malaria!); expeditions were sent from London just to bring back this plant.
Do this	• Scratch and sniff the twigs for a delightful spicy odor. • See if you can find all three leaf shapes. • Look for the odd dark blue fruit in a red cup on a red stem; if you find the fruit, you have also found a female tree. • Any Spicebush butterflies around, or their caterpillars?

For your sketches, notes, observations

	Shadbush or Serviceberry
	Amelanchier species
About the names	• <u>Shadbush</u> refers, apparently, to the time of bloom, which is also the time of the spawning run of shad fish.
	• <u>Serviceberry</u> is derived from a Latin word "sorbus," the name of a small pear tree similar to these species.
	• <u>Amelanchier</u> is a Provençal name, first recorded in 1741, for a member of this species.
	• Also called Juneberry, Wild Pear, Bilberry, Shadblow, Sarvis.
Memory aid	• Flowers (amid nearly bare branches) like a cloud of white foam caused by all those shad jumping around in the water while spawning.
When in bloom	• April-June; fruits June-September.
Interesting growth habits	• Flowers, described as five-bladed white propellers, bloom in earliest spring, often before leaves appear.
	• Early leaves may be folded along middle.
	• Winter buds are long, pointed, red and green or brown.
	• Ranges from 6-40 feet tall.
	• Leaves in fall turn yellow or orangey-red.
	• Fruits look like tiny purple apples.
Warnings, uses	• Fruits of all species are edible, some tastier than others; used raw or for jelly.
	• Native Americans mixed dried berries with meat to make pemmican, a rich food for traveling.
	• Fruit and twigs eaten by many birds and other animals.
"Stories"	• "A large, confusing genus."
	• "The Amelanchiers make up one of the 'difficult' groups within the Rose family. Whether there are few or many species depends upon the botanist consulted."
Do this	• In summer look to see if fruits are swollen and misshapen, and twigs covered with a powdery substance. These are marks of an infectious rust which needs both Shadbush and Red Cedar to complete its life cycle. If evidence of rust is found, look for the Red Cedars that must be near by.

For your sketches, notes, observations

POISONOUS	**Skunk Cabbage**
	Symplocarpus foetidus
About the names	• <u>Skunk Cabbage</u>: leaves smell bad and somewhat resemble those of a cabbage. • <u>Symplocarpus</u>: "fused fruit," referring to the tight cluster of fruits; <u>foetidus</u>: "bad-smelling."
Memory aid	• Like a skunk, the odd flowers and big leaves of this are a big presence in the woods; note that, then crush and sniff to confirm.
When in bloom	• February-April.
Interesting growth habits	• As the plant begins to grow in winter, it generates heat which may melt ice and snow around it, maintaining a temperature of about 72 degrees right around the flower. • First comes the thick purplish-brown spathe (hood) around the knobby spadix (flower cluster), followed by the unrolling of the big oval leaves from their pencil-shaped furl. • No stem on this plant; flowers and leaves arise directly from an underground rhizome. • The look and smell of the flower are that of rotting flesh, perfect to attract the flies that pollinate it.
Warnings, uses	• CAUTION Leaves and roots are toxic. • If roots are thoroughly dried, can be ground into flour. • Native Americans used for cramps, convulsions, whooping cough, toothaches, epileptic seizures.
"Stories"	• "[the Indians] smoak it when they want tobacco, and ere the knowledge of Rum was brought among them by the Christians they used to make a fuddling drink of it at their gambols and merry-making...I dryd and smoakt some on't but it stunk so wretchedly as to make me spew, but the Indians have a way of dressing it so as to make it less hideous." (1723)
Do this	• Crush a bit of leaf to experience the smell. • See if you can find some False Hellebore/Indian Poke to compare: they come out about the same time and in the same general habitat, but Hellebore has leaves before flowers, and its leaves clasp the stem and are elongated and oval rather than rounded and not clasping.

For your sketches, notes, observations

CAUTION	**Smartweeds**
	Polygonum species
About the names	• <u>Smartweed</u> may refer to the peppery taste of some species, or to the irritating quality of the juice. • <u>Polygonum</u> means "many knees," describing the swollen joints of the stems. • Common species are Pennsylvania Smartweed, Lady's-Thumb (with a dark print on each leaf), Water Smartweed, Pale Smartweed, Water Pepper (with extremely peppery taste), and two kinds of Tearthumb (with sharp prickles along veins).
Memory aid	• Smart enough to grow even in the drabbest places, such as the edges of sidewalks and parking lots.
When in bloom	• June-November.
Interesting growth habits	• Many species, found in dry, wet, and aquatic habitats. • Identification of different species often depends on the differing appearance of tiny hairs around the sheath found just above each stem joint. • What look like little pink or white petals are actually sepals. • Seeds may remain viable for a century or more, and germinate when soil is finally stirred up.
Warnings, uses	• CAUTION Juices may cause skin irritation. • Native Americans and others are reported to have used crushed Smartweeds, thrown into a pond, to stun and thus catch fish. • Contains a compound which helps stop bleeding. • At least thirty kinds of birds eat the seeds.
"Stories"	• Another one of those very common flowers that gets overlooked, but rewards a close look; once you identify it you see it everywhere. • Water Pepper was put under the saddle of a horse "to make him travel better," a practice dating back to the Scythians of 7th century B.C.
Do this	• If there are several plants together, pick one to examine with a magnifier not only the tiny flower (see if you can open it a bit), but also the stem joints and hairs. If you find a different species, compare them.

For your sketches, notes, observations

Solomon's Seal
Polygonatum biflorum

About the names	• <u>Solomon's Seal</u> refers to a round scar on the root stock; or possibly that wise King Solomon set his seal on the plant since it was so valuable as medicine. • <u>Polygonatum</u> means "many knees [joints]," describing the structure of the root stock; <u>biflorum</u> means "two flowers" since the flowers hang in pairs. • Also called Drop-Berry, Sealwort.
Memory aid	• 2 flowers, 2 berries, 2 S's in <u>S</u>olomon's <u>S</u>eal (and Se<u>SS</u>ile leaves [attached directly to stalk], unlike False Solomon's Seal).
When in bloom	• May-June.
Interesting growth habits	• Much less common than False Solomon's Seal (q.v.), but they may be found together. • Leaves may have disappeared when berries are ripe.
Warnings, uses	• Used in voodoo practices. • Much used for "sealing" wounds, possibly another explanation for its name. • Roots contain a chemical compound which has been used in modern medicine (from other sources) to treat wounds and skin ulcers. • Roots have been used like potatoes; young shoots, without leaf heads, cooked like asparagus; but berries are not edible.
"Stories"	• (About a similar European species, 1597) "The root of Solomons Seale stamped while it is fresh...taketh away in one night, or two at the most, any bruise, blacke or blew spots gotten by falls or womens wilfulnesses, in stumbling upon their hasty husbands fists, or such like." • "As a remedy for piles: 4 oz. Solomon's Seal, 2 pts. water, 1 pt. Molasses. Simmer down to 1 pt., strain, evaporate to the consistence of a thick fluid extract, and mix with it from ½ to 1 oz. Powdered resin. Dosage: 1 t. several times daily."
Do this	• Look under the leaves to find the pairs of dainty greenish-yellow flowers in spring and pairs of blue-black berries in summer.

For your sketches, notes, observations

| | **Common Speedwell** |
	Veronica officinalis
About the names	• <u>Speedwell</u>: Perhaps its little blue flowers were handed to travelers in the old days with the wish, "Speed well," or, alternatively, maybe because its use as a medicinal led to speedy healing. • <u>Veronica</u> means "true image" (the image of Christ, which the markings on this flower are thought to resemble, was left on a towel used by a woman, later St. Veronica, to wipe the sweat from his brow while he walked to Calvary); <u>officinalis</u> means "of the shops," probably because it was sold in apothecary shops as a medicine.
Memory aid	• Visualize plant as a traveler (it creeps along the ground) being wished "speed well."
When in bloom	• May-July or sometimes later.
Interesting growth habits	• Stem is hairy and creeps along the ground, sometimes forming mats. • Usually described as a weed; found in fields or open woods. • Found in Europe, British Isles, and Asia as well as North America.
Warnings, uses	• "Sodereth and healeth all fresh and old wounds, clenseth the bloud from all corruption, and is good to be drink for the kidneys, and against scurviness and foul spreading tetters [eczema, impetigo, etc.], and consuming and fretting sores, the smallpox and measels." (1633)
"Stories"	• Probably came over from England in ballast or seed.
Do this	• Look very closely to see the two tiny stamens and four 5 mm.-wide petals, the lower one shorter or smaller than the others.

For your sketches, notes, observations

	Spicebush
	Lindera benzoin
About the names	• <u>Spicebush</u>: a perfect name, for all parts of this bush smell deliciously spicy. • <u>Lindera</u>: for Johann Linder, an early Swedish botanist; <u>benzoin</u>: its smell is similar to Gum Benzoin, found in a native tree of the Far East.
Memory aid	• It's a bush, not a tree. Near water? scratch a twig and sniff: spicy? It's Spicebush.
When in bloom	• March-April; fruits July-September.
Interesting growth habits	• Clusters of flowers appear before leaves, from the axils of twigs; no petals, just sepals. • Very often found on stream banks.
Warnings, uses	• During the Revolution, Colonists substituted dried, powdered Spicebush fruits for allspice, being unable to get any from England. • During the Civil War, blockaded Southerners made tea from young bark, leaves, and twigs. • A favored food of wood thrushes and veeries • Host plant for Spicebush Swallowtail butterflies. • Numerous medicinal uses by Native Americans, including tea from leaves, bark, twigs, or berries, each part for a different ailment.
"Stories"	• Wonderful to see its puffs of yellow flowers amid the bare gray and brown branches of very early spring. • The chemicals in the plant that make its scent are hard to digest, a protection against herbivores; however, Spicebush Swallowtail's caterpillars have no trouble—though Spicebush is about all they eat (except for Sassafras, which has similar chemicals and scent).
Do this	• Watch for the lovely Spicebush Swallowtail butterfly, blue-black with orange spots, flying late spring and again late summer. Also look for its caterpillar, dark green with two conspicuous and fake "eyes"; it may lurk inside a folded-over leaf. • Scratch and sniff a twig; crush a leaf; smell a flower. • In winter, look for the dark green twigs with knobs of flower buds; are there any butterfly chrysalises on them?

For your sketches, notes, observations

Starflower
Trientalis borealis

About the names	• <u>Starflower</u>: both flowers and whorls of leaves below them are star-shaped.
	• <u>Trientalis</u> means "one-third foot," referring to its height of about four inches; <u>borealis</u> means "of the north," since it can be found at subalpine heights as well as lower.
	• Also called Star-Anemone, Star-Chickweed, Star-of-Bethlehem.
Memory aid	• A pair of pure white stars, sparkling with golden anthers, rise above a shiny green star of leaves.
When in bloom	• May-June.
Interesting growth habits	• Flowers have seven petals, seven stamens—a rare number among the wildflowers.
	• Stems are very delicate, like threads.
	• Indicates acid soil.
	• A "spring ephemeral," meaning that it blooms early, often before trees are fully leafed out, and then disappears, usually leaving no trace.
Warnings, uses	• Its only use is enjoyment of its elegant fragile beauty.
"Stories"	• "Some persons call this chick wintergreen, a name which is an insult to the plant...why, it is one of the daintiest wood-flowers, with nothing in the world to do with chicks, or weeds, or winter. It is not the least of an evergreen...and there is not a chicken in the country that knows it by sight or taste. Discriminating people, where they first find its elegant silvery flower growing in the woods beside the violet, call it May-star." (1850)
Do this	• Are plants really "trientalis"—a third of a foot tall?

For your sketches, notes, observations

| | **Staghorn Sumac** |
	Rhus typhina
About the names	• <u>Staghorn</u> refers to the appearance of its limbs; <u>Sumac</u>: "red" (Middle English via Old French via Medieval Latin via Arabic via Aramaic!). • <u>Rhus</u>: Ancient Greek name for the family; <u>typhina</u>: resembling cattails (the "furry" seed heads and branches of Sumac).
Memory aid	• Reddish-brown furry and forked branches like a stag's horns when in velvet, and with those big red ornaments on top—wow, <u>Su</u> the doe will certainly be won by <u>Mac</u> the stag!
When in bloom	• June–July; fruits June–September.
Interesting growth habits	• Upper branches covered with reddish-brown hairs. • Greenish-white flowers in a big upright cone, one sex to a plant; these become upright furry cones of red fruits, which persist in winter. • Twigs contain a milky sap.
Warnings, uses	• Soak the furry red fruit cluster in water after gently bruising the berries; strain (to remove insects and fur); sweeten; chill and drink: supposed to taste like pink lemonade, for it contains citric acid. • Leaves, berries, bark, and roots all used by Native Americans in various medications. • Used in leather tanning and dyeing. • By poking out the soft pith within a twig, a tube can be fashioned (called a spile) to serve as a spout and channel for Sugar Maple sap as it flows from holes in tree to buckets. • Many birds eat the fruits, and rabbits chew the bark in winter.
"Stories"	• Member of the Cashew family, as are Poison Ivy, pistachios, and mangoes. • "For gonorrhea: take 1 scruple [c. 1.3 grams] each of the exudation [from broken Sumac branches] and Canada balsam. Form into a pill mass with a sufficient quantity of powdered pokeroot, and divide into 10 pills, of which 1 or 2 may be taken three or four times daily."
Do this	• If you can find the flowers, see if their scent reminds you of cake batter, as noted by one naturalist; watch out for bees! • Examine the hairy fruits with a magnifier.

For your sketches, notes, observations

	Sweetfern
	Comptonia peregrina
About the names	• <u>Sweetfern</u>: looks like a fern (but isn't) and is spicy/sweet-smelling when crushed. • <u>Comptonia</u>: for Henry Compton, bishop of London and patron of botany; <u>peregrina</u>: "strange, foreign," possibly since it looks like a fern but isn't (most obviously because it has flowers and seeds, unlike true ferns).
Memory aid	• Name describes it perfectly.
When in bloom	• April-June; fruits September-October.
Interesting growth habits	• Not a fern at all but a shrub with woody stems. • A nitrogen-fixing plant, like beans and peas; bacteria live among its roots in tiny nodules and convert atmospheric nitrogen into a form usable by them and the plant. • Inconspicuous flowers in catkins show that it is wind-pollinated; they usually appear before the leaves, male and female on separate plants. • Brownish catkins persist in winter.
Warnings, uses	• Leaves can be made into a fragrant tea. • Native Americans used a wash of this plant as a remedy for poison ivy and other skin rashes.
"Stories"	• Sometimes called "Naughty Girls" because only one of them would roll in the Sweetfern (and you could probably tell she had been, if she walked by, Afterwards).
Do this	• Definitely crush and sniff the leaves. • Look on leaves with magnifier to find tiny yellow resin dots. • Look also for its bur-like fruits; open one to see the little nutlets.

For your sketches, notes, observations

POISONOUS	**Tansy**
	Tanacetum vulgare
About the names	• Both <u>Tansy</u> and <u>Tanacetum</u> mean "immortal," derived from Greek "athanatos" (without death), to Medieval Latin "athanasia" (an elixir of life), and from there to Old French "tanesie." • <u>Vulgare</u>: "common." • Also called Golden-Buttons, Bitter-Buttons.
Memory aid	• Tangy tansy, tansy's tangy (its strong odor).
When in bloom	• July-September (a warning that fall is coming).
Interesting growth habits	• Tansy has only disk flowers, with no "petals"; the flower head looks like a plate of mustard-yellow buttons. • Overwinters in a green rosette.
Warnings, uses	• CAUTION A strong dose of Tansy has been known to cause death; ½ oz. can cause death in 2-4 hours. • Used in the past to cause abortions, possibly resulting in death of both mother and child. • Used as a dye.
"Stories"	• Tansy was wrapped with the dead to deter insects. • In old times, Tansy was eaten in a cake at Lent, the bitter taste a symbol of Christ's suffering. The recipe for a tansy: "Beat 7 eggs, yolks and whites separately; add a pint of cream, near the same of spinach-juice, and a little tansy-juice gained by pounding in a stone mortar; a quarter of a pound of Naples biscuit, sugar to taste, a glass of white wine, and some nutmeg. Set all in a sauce-pan, just to thicken, over the fire; then put it into a dish, lined with paste, to turn out, and bake it."
Do this	• Gently bruise the leaves to release the pungent scent. • If it has gone to seed, rub the top of the seed head to loosen and examine the small seeds.

For your sketches, notes, observations

	Tick Trefoils
	Desmodium species
About the names	• Tick Trefoil: <u>Trefoil</u> means "three leaves" for the three-part leaflets; <u>Tick</u>: seeds stick like ticks on animals.
	• <u>Desmodium</u> means "chain," for the jointed seed pods.
	• May be called Sticktights, Beggar's-Lice, Devils'-Thistle.
Memory aid	• In fall, the seed pods almost seem to jump onto one's clothing, and s-<u>tick</u> like <u>ticks</u>.
When in bloom	• July-August; seed pods in September.
Interesting growth habits	• Member of the Legume (pea) family and thus host to a kind of bacteria, living in tiny nodules on the roots, which are able to convert nitrogen in the air to soil nitrogen, thereby fertilizing the soil around the plant.
	• Segmented pods will break apart, one seed in each part, when brushed by clothing or fur.
	• Seed pods are equipped with tiny curving hooks to help them stick.
	• All flowers of the Legume family have a distinctive shape: there are five petals, but the two lower ones may be joined in a kind of "keel" shape, the two side petals make "wings," and the top petal a "banner." This closed shape means that pollinating insects have to push into the flower, an activity which releases a kind of pollen trigger, shooting the pollen onto the insect's body.
Warnings, uses	• Check clothing (and dogs) for seed pods after a walk in the fall unless you want to introduce Tick Trefoils into your garden.
"Stories"	• "There is something witch-like about them; though so rare and remote, yet evidently, from those bur-like pods, expecting to come in contact with some travelling man or beast without their knowledge, to be transported to new hillsides; lying in wait, as it were, to catch by the hem of the berry-picker's garments and so get a lift to new quarters." (Thoreau, 1856)
Do this	• Examine seed pods under a magnifier.

For your sketches, notes, observations

Cow Vetch
Vicia cracca

About the names	• <u>Vetch</u> (as well as <u>Vicia</u>) comes from a Latin root meaning "to bend, to wind," referring to the vine-like nature of the plant (the "witch" in Witch Hazel comes from the same root); not clear whether cows eat it.
	• <u>Cracca:</u> an ancient word for Vetch.
	• Also called Blue Vetch, Bird Vetch, Tufted Vetch, Canada Pea.
Memory aid	• Vetch tends to lie around, prostrate, sprawling, leaning on or around other plants, obviously feeling wretched: <u>wretch</u>ed <u>V</u>etch.
	• Also, it is considered a nasty weed; a wretch to have Vetch in your garden.
When in bloom	• May-August.
Interesting growth habits	• A legume: nodules on the roots are home to bacteria which are able to capture nitrogen gas, which is then converted to a form usable by plants.
	• The only one of the common Vetches which has flowers on only one side of the flower stalk.
	• Fruits look just like tiny pea pods (no surprise since peas are also legumes) that turn light brown when mature.
Warnings, uses	• Vetch can easily climb up landscape and crop plants, so best not to allow it in the garden.
"Stories"	• Although this plant is found all over, it's spurned when it comes to writing about it! Attention should be paid; it's an opportunistic plant with wonderful complicated "pea-type" flowers.
Do this	• Check to see if pods are present; open one to see the little "peas." If it's fall, and the pods are dry, gently shake the plant to see if the dry peas rattle inside their pods.
	• Use a magnifier to examine the blue-purple flowers.
	• Look for the tiny bristle tips on the end of the downy leaflets.

For your sketches, notes, observations

Maple-Leaved Viburnum
Viburnum acerifolium

About the names	• <u>Maple-Leaved</u>: it is. • <u>Viburnum</u>: Latin name for Wayfaring-Man's-Tree, a member of this genus; <u>acerifolium</u> means "maple-leaved." • Also called Dockmackie.
Memory aid	• If its leaves are opposite and it's a shrub, not a tree, and its leaves look like those of a Maple, why, it's a Maple-Leaved Viburnum.
When in bloom	• May-August; fruits July-October.
Interesting growth habits	• Leaves turn a distinctive pink-magenta in fall. • Only about four feet tall. • Fruits held high on stiff upstanding stalks. • This looks a bit like another Viburnum (called Highbush Cranberry), but the latter's flower clusters are surrounded by some "flowers" which are large and showy but sterile, with no sexual parts; Maple-Leaved Viburnum's flower clusters contain all fertile flowers.
Warnings, uses	• Native Americans used a tea made of the bark as an emetic.
"Stories"	• The fruits have very little fat (it is "expensive" for the plant to produce it), so animals eat them late in the winter, after other, fattier fruits are taken. • One study showed over 70% of the fruit was still on the bush on January first.
Do this	• Feel the velvety twigs and downy leaves, which have tiny yellow and black dots below. • Look closely at the leaves to see the pair of tiny leaf-like stipules where they attach to the twigs. • In spring and summer, see if you can find rose-colored caterpillars with accompanying ants, the larvae of Spring Azure butterflies.

For your sketches, notes, observations

POISONOUS	**Virginia Creeper**
	Parthenocissus quinquefolia

About the names	• <u>Virginia</u> does not appear to refer to the st ate, but rather to the Latin name (see below); <u>Creeper</u>, since it is a rapidly-spreading vine
	• <u>Parthenocissus</u>: "virgin-ivy"; <u>quinquefolia</u>: "five-leaved."
	• Also called Woodbine.
Memory aid	• "Leaves of five, stay alive" 5 = V [Roman numeral] = Virginia.
When in bloom	• June-August; fruit August-February.
Interesting growth habits	• Climbs by means of tendrils with adhesive disks at their ends.
	• Spreads by seeds dropped by birds, but also by stems, which root when they touch the ground.
	• Adhesive disks form only after tendril has touched something.
	• One of first plants to turn red in fall.
Warnings, uses	• CAUTION Berries can be fatal if eaten.
	• CAUTION Touching leaves may cause skin irritation in some people.
	• High-fat fruit is attractive to many birds, especially when migrating; a special favorite of pileated woodpeckers.
"Stories"	• A group of five tendrils can hold ten pounds of weight!
	• It's said that old fruits may ferment, and if eaten may lead to intoxicated birds.
Do this	• Look to see how tendrils appear opposi te leaf stalks.
	• Look for clusters of small greenish -white flowers, or blue-black berries.
	• See if you can find some Poison Ivy nearby, with its "leaves of three," to compare.
	• Several species of sphinx moths eat this plant as their main food; look closely to see if their caterpillars are present.

For your sketches, notes, observations

| | **Fragrant Water Lily** |
	Nymphaea odorata
About the names	• <u>Fragrant Water Lily</u>: an apt name. • <u>Nymphaea</u>: for the Greek deities called Nymphs, who were beautiful and lived in water, just like this flower; <u>odorata</u> means "scented." • Called Alligator-Blankets in the South.
Memory aid	• Everyone knows the classic water-lily flower.
When in bloom	• June-September.
Interesting growth habits	• Leaves are purplish underneath; the dark color is thought to raise the temperature of the leaf a little above that of the water, an arrangement which helps transpiration. • Stems contain tubes for transport of air and gases back and forth (see Yellow Pond Lily). • Stomata (tiny openings) for gas exchange are on upper surface of leaf, rather than below leaf as is true for land-dwelling plants. • Flowers open only on sunny days, and close at noon. • After fertilization, the flower closes and the stem curls down, carrying the flower with it; under water the seeds mature in a container which then breaks off and floats to the surface, where the seeds are released. • A first-year plant is under water; the second year it makes leaves only; by the third year, it finally flowers.
Warnings, uses	• Young leaves and unopened flower buds can be boiled and served like vegetables; seeds are protein-rich and can be dried and cooked like popcorn or made into flour. • Contains mucilaginous substance used to soothe sore throats.
"Stories"	• Used to be prescribed as an anaphrodisiac, to inhibit sexual drive; it's suggested that the chaste-appearing flower rising out of the muck may have indicated such a use. Thoreau said, "How pure its white petals, though its root is in the mud!"
Do this	• If you can get close to a flower somehow, without falling in, smell it! • If you can reach some leaves, turn them over to see what evidence of creatures you can find.

For your sketches, notes, observations

POISONOUS	**Winterberry**
	Ilex verticillata

About the names	• <u>Winterberry</u>: bright red berries very prominent and showy during winter. • <u>Ilex</u>: Latin name for an evergreen Oak; since most of the genus Ilex (Holly) is evergreen (though not this plant), name was applied to it; <u>verticillata</u> means "arranged in whorls," which is a mystery since nothing on this plant seems to have that characteristic. • Also called Christmasberry, Redberry, and, confusingly, Black Alder.
Memory aid	• Easy in winter: brilliant and abundant winter berries make a great show, nestling close to the stem, as if to keep warm.
When in bloom	• June-August; fruits September-March.
Interesting growth habits	• Leaves, taken as a whole on the plant, have a kind of regular, orderly appearance. • Some plants are male, and have no berries; others are female, with berries, typical of Holly family. • Can grow to be size of small tree—6-15 feet. • Unlike some other Hollies, it sheds its leaves, which turn dark brown or black before falling.
Warnings, uses	• CAUTION: Berries are poisonous! • Native Americans made an astringent medication from the bark. • Many birds are attracted to and eat the berries. • Bare branches with clusters of red berries used for Christmas decorations.
"Stories"	• Thoreau observed that mice run up the trunk of the Winterberry, at night, to gather the fruits, to get at the seeds within them. • However, the fruits are low in fat, and so are not especially rewarding nutritionally speaking; animals will eat them only after fruits with more fat are gone (one reason the berries persist so long into the winter). It is "expensive" for a plant to produce fats.
Do this	• If you find a female plant, with berries, look around to find the male, without. There has to be one, since the male's pollen must fertilize the female in order to produce the fruit.

For your sketches, notes, observations

CAUTION	**Wintergreen**
	Gaultheria procumbens

About the names	• <u>Wintergreen</u> describes it perfectly. • <u>Gaultheria</u> for Dr. Gaulthier, a Canadian botanist of the 18th century; <u>procumbens</u> means "lying flat on the ground, creepin g forwards" • Also called Teaberry, Checkerberry; or Boxberry, Youngsters, Gingerplum, Pippins, Ivory-leaves, Jinks, Ivory-Plums, Young-Plantlets, Drunkards, Foxberry...and on and on!
Memory aid	• One sniff of a crushed leaf of this evergreen plant, and you'll know.
When in bloom	• July-August.
Interesting growth habits	• A shiny-leaved evergreen (actually not green in winter, but rather a brownish green). • Indicates acid soil, paucity of nutrients. • It's actually a tiny shrub; the growth you see is twigs that grow from the creeping stem, which is usually underground.
Warnings, uses	• CAUTION Large quantities of the oil are toxic, absorbed through the skin. • Some of you will be old enough to remember Teaberry chewing gum—was it in a pink package? • Make Wintergreen tea from the dried leaves. • Contains an aspirin-related compound, so used for headaches, rheumatism.
"Stories"	• During the tea boycott of the American Revolution, the patriotic Colonists drank wintergreen tea. • Chemically, Wintergreen oil is very similar to the oil in Black and Yellow Birches.
Do this	• Crush and sniff a leaf; biting into one is good, too, but it doesn't taste as good as it smells. • If you are lucky enough to see the flowers, use a magnifier to enjoy their dainty bells. • Can you find the red fruits in winter? If there are plenty of them, try tasting one, and look for the star shape on one end.

For your sketches, notes, observations

| | **Witch Hazel** |
	Hamamelis virginiana
About the names	• <u>Witch</u> comes from an Anglo-Saxon word <u>wych</u>, meaning "to bend" ("wicker" also derived from this word), referring to the use of its twigs as divining rods; <u>Hazel</u> due to its supposed resemblance to a Hazel shrub. • <u>Hamamelis</u>: Greek for a plant with pear-shaped fruits; <u>virginiana</u>: found in Virginia.
Memory aid	• Witch Hazel has blasted these trees! See how irregular and ratty-looking the leaves are—why, even their bases are irregular, with one lobe bigger than the other. It's from the spell she laid on the leaves. She leaves pointy little witch hats (galls) on the leaves. There's her stringy yellow hair, too (flowers), when the tree is in bloom.
When in bloom	• Sept.-Nov./Dec.; fruits Aug.-Oct.
Interesting growth habits	• The last plant to flower, it takes advantage of whatever hungry pollinators are around. • Seeds take a year to mature in their 4-parted pea-sized capsules, which in late fall dry, contract and shoot out seeds (5-50 feet depending on your source of information) which then remain dormant for two winters. • Long, dangling flower petals can recurl into a bud shape when the temperature drops, and reopen when warmer.
Warnings, uses	• Witch Hazel oil, from bark and twigs, was and is used extensively in a soothing, astringent lotion for many afflictions; it is still available at pharmacies. • Twigs used as divining rods, mysteriously useful in finding water, coal, tin, copper, lost objects.
"Stories"	• Two interesting galls (growths caused by insect eggs laid within a plant) on this tree: one, on leaves, shaped like a tiny witch hat, caused by the conegall aphid; the other, on twigs, looks like a spiny little pineapple, caused by the spiny witch-hazel budgall aphid (both aphids spend remainder of their life cycle on White Birches). • Leaves also home to two leaf engineering insects, one which ties and one which rolls the leaves to make homes for larvae.
Do this	• Look for: empty seed pods, pointy and spiny galls, rolled or tied leaves.

For your sketches, notes, observations

	Witherod *Viburnum cassinoides*
About the names	• <u>Witherod</u>: stems are lithe and flexible; comes from Latin word for "vine." • <u>Viburnum</u>: Latin word for Wayfaring-Man's-Tree, a member of this genus; <u>cassinoides</u>: resembling *Ilex cassine*, a kind of Holly. • Also called Wild Raisin.
Memory aid	• To remember it as Witherod: see long, flexible, slender twigs. • To remember it as Wild Raisin: leaf shapes are highly variable, just like individual raisins.
When in bloom	• June-August; fruits September-October.
Interesting growth habits	• Leaves variable, sometimes almost toothed, sometimes wavy-edged, sometimes almost smooth. • In spring, new leaves are bronze-colored. • Fruits begin as bright pink, ripen to dark blue. • Stalked clusters of sweet-smelling tiny white flowers.
Warnings, uses	• Flexible stems were used like cord. • Sweet fruit can be eaten by people and birds. • Leaves have been used to make tea.
"Stories"	• In some years, may bear prolifically, making a brilliant, rich show of fruit. • If not eaten, fruit tends to shrivel on the stalk, thus leading to the name of Wild Raisin. • Donald Stokes, on the buds opening in spring: "First, the long rough scales are shed and pairs of leaves open in perfect unison. Their sides are folded up along the midvein, their exterior is reddish and granular like the bud scales, and their finely cut teeth look like jewels along the edge of the leaf when seen with the light behind them. Framed by the graceful lines of the opposite twigs, these features create a curving symmetry. Visit these plants in this season and see if you don't agree."
Do this	• In spring, look for large flask-shaped flower buds with pair of rough, yellowish scales not quite covering the bud. • Open the fruit to see the large flat seed.

For your sketches, notes, observations

POISONOUS	**Yarrow**
	Achillea millefolium
About the names	• <u>Yarrow</u> is an Old English name, the derivation of which is uncertain.
	• <u>Achillea</u> refers to the Greek hero Achilles, who is supposed to have used this plant to stem the flow of blood from his soldiers' battle wounds; <u>millefolium</u> describes the thousand leaves the Yarrow appears to have.
	• Also called Bloodwort, Milfoil, Stanchgrass, Nosebleed-Plant.
Memory aid	• Feathery, lacy leaves look like those of the carrot (same family); think c<u>arrot</u>/y<u>arrow</u>.
When in bloom	• June-September.
Interesting growth habits	• Forms a pretty ground-hugging green rosette in winter.
Warnings, uses	• CAUTION May cause skin irritation.
	• CAUTION Contains a toxic compound.
	• When building models, architects use the dried flower stalks to represent trees; no matter how short they are cut, the "branches" still loo k treelike.
	• Used for generations by many cultures to treat numerous ailments from hemorrhoids to baldness to rheumatism to the common cold.
	• Contains 100+ biologically active compounds.
"Stories"	• Called Soldier's-Woundwort during the Civil War, for its use then.
	• If you stick a leaf in your nose, turn it around three times (the leaf, not the nose), and your nose bleeds, you will win your love.
	• Zuni Indian fire handlers reported to have used Yarrow topically and internally before handling hot coals.
	• Brought over by early settlers to plant as medicinal.
	• The best story: remnants of Yarrow were found in a 60,000-year old Neanderthal burial, in Iraq.
Do this	• Crush leaves for distinctive medicinal scent.
	• Use a magnifier to examine tiny flowers.

For your sketches, notes, observations

BIBLIOGRAPHY

I am deeply indebted to Timothy Coffey, John Eastman, and Donald Stokes, not only for their painstaking and clever research, upon which I have drawn, but even more importantly for the inspiration I have received from their loving, enthusiastic attention to the awesome lives of plants.

Alden, Peter et al. *National Audubon Society Field Guide to New England.* New York: Alfred A. Knopf, 1998.

Baldwin, Henry Ives. *Forest Leaves: How to Identify Trees and Shrubs of Northern New England.* Portsmouth, New Hampshire: Peter E. Randall Publisher, 1982.

Blouin, Glen. *Weeds of the Woods: Small Trees and Shrubs of the Eastern Forest.* Fredericton, New Brunswick, Canada: Goose Lane Editions, 1992.

Brown, Lauren. *Weeds in Winter.* New York: W. W. Norton and Company, 1976.

Carter, Gale Winston. *Let's Take A Walk.* Colebrook, New Hampshire: Liebl Printing Company, 1991.

Cobb, Broughton. *A Field Guide to Ferns and Their Related Families.* Boston: Houghton Mifflin Company, 1956.

Coffey, Timothy. *The History and Folklore of North American Wildflowers.* Boston: Houghton Mifflin Company, 1993.

Eastman, John. *Forest and Thicket.* Harrisburg, Pennsylvania: Stackpole Books, 1992.

----. *Swamp and Bog.* Mechanicsburg, Pennsylvania: Stackpole Books, 1995.

Embertson, Jane. *Pods: Wildflowers and Weeds in Their Final Beauty.* New York: Charles Scribner's Sons, 1979.

Foster, Steven and Roger Caras. *A Field Guide to Venomous Animals and Poisonous Plants.* Boston: Houghton Mifflin Company, 1994.

Foster, Steven and James A. Duke. *A Field Guide to Medicinal Plants.* Boston: Houghton Mifflin Company, 1990.

Gledhill, D. *The Names of Plants.* Cambridge, England: Cambridge University Press, 1989, 2nd ed.

Grieve, Mrs. M. *A Modern Herbal.* 2 vols. New York: Dover Publications, Inc., 1971. (Originally published by Harcourt, Brace and Company, 1931.)

Grimm, William Carey. *The Illustrated Book of Wildflowers and Shrubs.* Mechanicsburg, Pennsylvania: Stackpole Books, 1993, rev. ed.

Harris, James G. and Melinda Woolf Harris. *Plant Identification Terminology.* Spring Lake, Utah: Spring Lake Publishing, 1994.

Heywood, V. H., ed. *Flowering Plants of the World.* New York: Oxford University Press, 1993.

Jorgensen, Neil. *A Sierra Club Naturalist's Guide: Southern New England.* San Francisco: Sierra Club Books, 1978.

Kricher, John C. and Gordon Morrison. *A Field Guide to Ecology of Eastern Forests.* Boston: Houghton Mifflin Company: 1988.

Levine, Carol. *A Guide to Wildflowers in Winter.* New Haven, Connecticut: Yale University Press, 1995.

Little, Elbert L. Jr. *Forest Trees of the United States and Canada, and How to Identify Them.* New York: Dover Publications Inc., 1979. (First published by U. S. Forest Service in 1978.)

Magee, Dennis W. *Freshwater Wetlands: A Guide to Common Indicator Plants of the Northeast.* Amherst, Massachusetts: The University of Massachusetts Press, 1981.

Magic and Medicine of Plants. Pleasantville, New York: The Reader's Digest Association, 1986.

Montgomery, Sy. *Nature's Everyday Mysteries.* Shelburne, Vermont: Chapters Publishing Ltd., 1993.

Newcomb, Lawrence. *Newcomb's Wildflower Guide.* Boston: Little, Brown and Company, 1977.

Niering, William A. and Nancy C. Olmstead. *National Audubon Society Field Guide to North American Wildflowers (Eastern Region).* New York: Alfred A. Knopf, 1979.

Peterson, Lee Allen. *A Field Guide to Edible Wild Plants.* Boston: Houghton Mifflin Company, 1977.

Peterson, Roger Tory and Margaret McKenny. *A Field Guide to Wildflowers.* Boston: Houghton Mifflin Company, 1968.

254

Petrides, George A. *A Field Guide to Trees and Shrubs*. Boston: Houghton Mifflin Company, 1972, 2 nd ed.

Prance, Ghillean T. Paintings by Anna Vojtech. *Wildflowers for All Seasons*. New York: The New York Botanical Garden Press, 1989.

Saunders, Gary. *Trees of Nova Scotia*. Halifax, Nova Scotia, Canada: Nimbus Publishing Limited and Province of Nova Scotia, 1970, 3 rd ed.

Scott, Jane. *Botany in the Field*. Englewood Cliffs, New Jersey: Prentice-Hall, Inc., 1984.

Slack, Nancy G. and Allison W. Bell. *AMC Field Guide to the New England Alpine Summits*. Boston: Appalachian Mountain Club Books, 1995.

Steele, Frederic L. *At Timberline: A Nature Guide to the Mountains of the Northeast*. Boston: Appalachian Mountain Club, 1982 .

Stokes, Donald W. *The Natural History of Wild Shrubs and Vines*. New York: Harper and Row, Publishers, 1981.

----. *A Guide to Nature in Winter*. Boston: Little, Brown and Company, 1976.

Stokes, Donald and Lillian. *The Wildflower Book*. Boston: Little, Brown and Company, 1992.

Symonds, George W. D. *The Tree Identification Book*. New York: William Morrow and Company, 1958.

----. *The Shrub Identification Book*. New York: William Morrow and Company, 1963.

Uva, Richard H., Joseph C. Neal and Joseph M. DiTomaso. *Weeds of the Northeast*. Ithaca, New York: Comstock Publishing Associates of Cornell University Press, 1997.

Wallner, Jeff and Mario J. DiGregorio. *New England's Mountain Flowers*. Missoula, Montana: Mountain Press Publishing Company, 1997.

Wherry, Edgar T. *The Fern Guide*. New York: Dover Publications Inc., 1995. (First published by Doubleday and Company in 1961.)

Wiley, Farida A. *Ferns of Northeastern United States*. New York: Dover Publications, Inc., 1973. (First published privat ely in 1948.)

Index

T

U

V

W

Y

ORDERING INFORMATION

If you would like to order additional copies of **Never Say It's Just A Dandelion**, please send a check for $12.00 (includes $2 for shipping) for each copy, plus the completed shipping form below, to:

Jewelweed Books
30 Winslow St.
Cambridge, MA 02138

MA residents please add $.50 tax for each copy.

Ship To:

Name: _____

Address: _____

_____

_____

Email: _____

Number of copies ordered: _____

Amount Enclosed: $ _____

Thank your for your interest.

JewelweedBooks@aol.com